STUDENT STUDY GUIDE

RICHARD F. JONES
JOHN W. HILL

JOHN W. HILL
DORIS K. KOLB

CHEMISTRY
for Changing Times
NINTH EDITION

Prentice Hall

Upper Saddle River, NJ 07458
http://www.prenhall.com

Executive Editor: Kent Porter Hamann
Project Manager: Kristen Kaiser
Special Projects Manager: Barbara A. Murray
Production Editor: Dinah Thong
Supplement Cover Manager: Paul Gourhan
Supplement Cover Designer: PM Workshop Inc.
Manufacturing Manager: Trudy Pisciotti
Cover Photo: Peter Beck/The Stock Market

Printed in the United States of America

10 9 8 7 6 5 4 3 2 1

ISBN 0-13-087497-3

Prentice-Hall International (UK) Limited, London
Prentice-Hall of Australia Pty. Limited, Sydney
Prentice-Hall Canada, Inc., Toronto
Prentice-Hall Hispanoamericana, S.A., Mexico
Prentice-Hall of India Private Limited, New Delhi
Pearson Education Asia Pte. Ltd., Singapore
Prentice-Hall of Japan, Inc., Tokyo
Editora Prentice-Hall do Brazil, Ltda., Rio de Janeiro

Contents

How to Survive in Chemistry

With a Satisfactory Grade

There is nothing worse than the feeling of frustration that you experience when faced with the will to succeed with subject matter that seems difficult. There are, however, a few things that you can do to help yourself. For instance, you have a chemistry textbook. Read the appropriate chapter before you attend class so that you'll have a general idea of what the teacher will be covering. After class, review your notes and reread the sections of the textbook that help explain any parts of your notes you don't understand. Try to relate one topic to another. The different topics should seem integrated to you. Go over the practice problems in the chapter and complete the problems at the end of the chapter. There's a slogan: "CHEMISTRY IS NOT A SPECTATOR SPORT." This means that you have to get off the sidelines and be able to apply yourself to chemistry. Success in chemistry, like success in sports or music, requires practice. If you need help completing the problems, ask for it. Your teacher will be able to help you.

Finally, as an aid to learning chemistry, you have this study guide. It provides a list of terms, a summary (in outline form), a list of objectives, and a self-test with answers for each chapter. Where appropriate, a discussion of the material in the chapter is also provided. Please note that there isn't a self-test item for every objective or an objective for every test item, but the self-test questions should provide you with an indication of how well you understand the material. For some chapters, additional worked-out examples are provided and additional problems are included. You should work those problems when they or similar ones from the textbook are assigned.

Your teacher and the textbook can be of considerable help in learning chemistry, but your success in the course will depend mainly on what *you* do. Be sure, before each test, that you have completed all assignments. Use the self-test to be sure that you understand the material in each chapter.

We hope that you find this study guide helpful. Please send us any criticisms and suggestions for improvement. A notation of any errors that you find would be especially helpful.

John W. Hill
Department of Chemistry
University of Wisconsin
River Falls, WI 54022

Richard F. Jones
Department of Chemistry
Sinclair Community College
Dayton, OH 45402-1460

Study Strategy Suggestions

There are many effective ways to study but most of them incorporate some of the following:

1. Preview the material in the text to be covered in the lecture.

    Just 10 minutes of leafing through the pages can make the lecture more meaningful and your note-taking easier.

2. Within two hours of the lecture, quickly read through the notes from the last lecture.

    You forget most of what you hear within two hours. Ten minutes of refreshing your memory shortly after the lecture is a very effective way to retain information.

3. For each hour of lecture, you should spend a minimum of three hours of study.

4. After reading the text and studying the lecture notes, move quickly to problems at the end of the chapter and the self-tests in the study guide.

5. Four ½-hour study periods are better than two 1-hour study periods, which is better than one 2-hour study period.

6. If possible, study with a friend and/or a study group.

    Each member of an effective study group should prepare individually for the study session. The strengths of each person will help eliminate the weaknesses of the others.

7. Prepare for the exams well in advance. Arrive early and have all materials needed for the test.

Chemistry

A Science for All Seasons

KEY TERMS

alchemy	heat	potential energy
applied research	hypotheses	risk-benefit analysis
atom	Kelvin	science
basic research	kilocalorie	scientific law
calorie	kilogram	scientific model
Celsius scale	kinetic energy	SI units
chemical change	liquid	solid
chemical property	liter	substance
chemical symbol	mass	technology
chemistry	matter	temperature
compound	meter	theory
density	mixture	variable
element	molecule	weight
energy	physical change	
gas	physical property	

Chapter Summary

1.1 Science and Technology: the Roots of Knowledge
    A.   Knowledge may be technological (factual) or philosophical (theoretical).
    B.   The ancient Greeks formulated theories about nature.
    C.   Alchemy tried unsuccessfully to turn metal into gold.
        1. Alchemy perfected some chemical techniques such as distillation and extraction.
    D.   Science developed out of natural philosophy but had its true beginnings when people began to rely on experiments.

1.2 The Baconian Dream and the Carsonian Nightmare
    A.   Francis Bacon envisioned that science would bring new inventions and prosperity.
    B.   Bacon's dream appeared near fulfillment by the middle of the twentieth century.

C.   Rachel Carson predicted disaster from the use of chemicals to control insects.

D.   Malthus earlier had predicted that the population would outstrip the food supply with famine as a consequence.

1.3   Science: Testable, Reproducible, Explanatory, and Tentative

A.   Scientists formulate testable hypotheses.

B.   Data must be reproducible.

C.   Scientific facts are verified by testing.

D.   Scientists observe carefully and measure accurately.

E.   Scientific laws summarize large amounts of data.

    1. Scientific laws can often be stated mathematically.

F.   Correlation between data does not necessarily prove cause and effect.

G.   Scientific theories are used to explain the behavior of matter.

H.   Scientific models help us to visualize invisible processes.

I.   Science is testable, explanatory, and tentative to the establishment of cause and effect.

J.   Computers can generate 3–dimensional representations of molecules giving a lot of chemical information.

1.4   The Limitations of Science

A.   Scientists try to control variables in an experiment.

B.   It is difficult to control variables in social experiments.

    1. Social scientists use some of the methods of scientists.

1.5   Science and Technology: Risk and Benefits

A.   Technology is the sum of the processes by which we modify materials to better serve us.

B.   Risk-benefit analysis involves the calculation of a desirability quotient:

$$DQ = benefits/risks$$

C.   Benefits and risks often are difficult to quantify, leading to uncertain DQs.

1.6   Chemistry: Its Central Role

A.   Chemistry is important to the other sciences and to social goals.

B.   Chemistry is important to the economy.

    1. Chemical trade helps lower the United States trade deficit.

1.7   Solving Society's Problems: Scientific Research

A.   Many chemists are involved in applied research.

B.   Applied research is directed toward the solution of a particular problem in industry or the environment.

C.   Many chemists are engaged in basic research.

D.   Basic research is the pursuit of knowledge for its own sake.

1.8   Chemistry: A Study of Matter and its Changes

A.   Chemistry deals with matter and the changes that it undergoes.

B.   Matter occupies space and has mass.

    1. Mass measures a quantity of matter that is independent of its relative location.

    2. Weight measures a force, such as the gravitational force or attraction between an object and the Earth.

C. Matter is characterized by its properties.
   1. Chemical properties describe and determine how substances combine with other substances.
   2. Chemical change results in a change in chemical properties.
   3. Physical properties can be observed and specified without reference to any other substance.
   4. Physical change does not change the composition or chemical nature of a substance.

1.9   Classification of Matter
  A. The States of Matter
     1. Solids maintain shape and volume.
     2. Liquids maintain volume but not shape.
     3. Gases maintain neither volume nor shape.
  B. Matter: Pure Substances and Mixtures
     1. Pure substances have a fixed composition.
     2. A mixture is a collection of two or more substances that can be mixed in any proportion because they are not chemically bonded.
  C. Elements and Compounds
     1. An element is a pure substance that is defined by a single type of atom.
       a. At present there are 115 known elements.
       b. Each element is represented by a chemical symbol.
     2. A compound is made up of two or more elements and has a fixed composition.
     3. Chemical symbols are made up of one, two, or three letters derived from the English (or Latin) name of the element.
       a. Only the first letter of a chemical symbol is capitalized.
     4. Atoms and Molecules
       a. Atoms are the smallest characteristic part of an element.
       b. Molecules are the smallest characteristic part of a compound.

1.10 The Measurement of Matter
  A. SI is an updated metric plan.
  B. Length-meter (m)—slightly more than a yard
  C. Mass-kilogram (kg)—(2.2 lb)
  D. Volume-liter (L)—slightly more than a quart
     1.1 mL = 1cm^3
  E. The commonly used metric prefixes are:
     1. kilo- (k) x 1000.   (10^3)
     2. deci- (d) x 0.1   (10^{-1})
     3. centi- (c) x 0.01   (10^{-2})
     4. milli- (m) x 0.001   (10^{-3})
     5. micro- (μ) x 0.000001 (10^{-6})

1.11 Density is mass per unit volume.
   A. $D = m/v$.
   B.   Density is usually expressed in units of grams per milliliter (g/mL) or grams per cubic centimeter (g/cm^3).
   C.   The density of water is 1.0 g/mL.

1.12 Energy: A Matter of Moving Matter
   A.   Energy is the ability to change matter.
      1.   Potential energy is that energy due to an object's position or arrangement.
      2.   Kinetic energy is the energy of motion.
      3.   Nearly all energy on Earth comes ultimately from the sun.
   B.   Definition of Heat and Temperature
      1. Heat is the measure of how much energy a sample contains.
      2. Temperature is the measure of average energy of the particle of a sample.
   C.   Energy-joule (J) or calorie (cal)
         1.0 cal = 4.18 J; 1 Kcal = 1 Calorie (foods) = 1000 cal
      1.   Temperature and Heat
         a.   Celsius scale (°C). Water freezes at 0°C and boils at 100° C.
         b.   Kelvin scale (K)—absolute scale. 0 K = –273°C.
         c.   Conversion between Celsius and Kelvin scales K = °C + 273

1.13 Critical Thinking: FLaReS
      1.   Falsibility: Can you prove something wrong?
      2.   Logic: Does conclusion follow from premises; are premises true?
      3.   Replicability: If experiment is cited, can experiment be repeated to produce same evidence?
      4.   Sufficiency: Evidence provided must be adequate.
         a.   Burden of proof on claimant
         b.   Extraordinary claims require extraordinary evidence.
         c.   Evidence based on authority and/or testimony is not adequate.

CHAPTER OBJECTIVES

These objectives are a minimal list. In addition to knowing the material specifically indicated here, you should do the assigned problems in the text. Your instructor may provide additional (or alternative) objectives for any chapter.

(Each statement in the following list is preceded by "You should...")

1.   Be able to define or identify each of the key terms.

2.   Know that Sir Francis Bacon was the first person to predict that science would endow us with new inventions and wealth.

3.   Know that the main theme of Rachel Carson's *Silent Spring* was that we might destroy all life on Earth through excessive use of chemical pesticides.

4. Know that four characteristics of science are that it is testable, reproducible, explanatory, and tentative.

5. Know that the reproducible experiment is the characteristic that most clearly sets science apart from the other disciplines.

6. Know that physical scientists (physicists, chemists) usually have the most control over variables in their experiments, social scientists the least (biologists are somewhere in between).

7. Be able to estimate a desirability quotient, given an estimate of risks and weight.

8. Be able to distinguish between mass and weight.

9. Be able to distinguish between chemical and physical properties.

10. Be able to distinguish between chemical and physical change.

11. Be able to distinguish between kinetic energy and potential energy.

12. Know that gases retain neither shape nor volume.

13. Know that liquids retain volume but not shape.

14. Know that solids retain both shape and volume.

15. Be able to distinguish between pure substances and mixtures.

16. Be able to give the symbol of each of the elements in Table 1.1 of the text, provided you are given the name.

17. Be able to give the name (with correct spelling) of each of the elements in Table 1.1 of the text, provided you are given the symbol.

18. Be able to convert units within the metric system (e.g., kg to g, mL to L, etc.).

19. Be able to calculate density, mass, or volume, given the other two quantities.

20. Be able to convert °C to K and vice versa.

21. Know that chemistry is fundamental to other sciences and society's goals.

22. Know that basic research is the pursuit of knowledge for its own sake.

23. Know that applied research is directed toward immediate, practical results.

24. Be able to apply the "FLaReS" critical thinking test.

DISCUSSION

Much of Chapter 1 of the textbook is intended to place chemistry in both historical and contemporary perspective—to give you a feeling for chemistry as it affects society. If, after reading the chapter, you recognize chemistry as something more than just a course that meets certain requirements, you have indeed understood what we were trying to say. In addition to this overview, Chapter 1 introduces several concepts important to our further study of chemistry. These include the international system of measurement, the meaning of terms such as matter and energy, different temperature scales, chemical symbols, and density. The problems at the end of the chapter are meant to check your understanding of this material. The following questions offer another opportunity for you to test yourself on Chapter 1.

Note: Many of these problems are more than tests of your memory. A number of them require preliminary calculations before an answer can be selected. You are expected to know the metric prefixes and units of measure, but consult **Appendix A** in the text if you need help with the metric conversion factors.

EXAMPLE PROBLEM

Challenge: Name something that is not a chemical.

Answer: Everything is a chemical. Energy and matter are equivalent. A vacuum is the absence of matter, which begs the question: It is nothing. Also a complete vacuum does not exist even in space. Some other possible responses: An idea (Can a thought exist without a brain with neurons and electrons?) Time (Can time exist without something around to mark its passing?) Basically chemistry is the study of just about everything.

ADDITIONAL PROBLEMS

1. An ad on TV invites you to call for a free "psychic reading." As proof of the psychics legitimacy they offer several testimonials by people claiming that the psychic they called knew facts about the caller that were known by only themselves. Which part of the "FLaReS" test does the ad fail?

    a. Falsibility
    c. Replicability

    b. Logic
    d. Sufficiency

    It fails all parts of the test:

    a. Falsibility—can't ask testimonial givers questions
    b. Logic—not applicable
    c. Replicability—can't reproduce results
    d. Sufficiency—no evidence given

2. A horoscope printed in the newspaper {*Dayton Daily News*; Joyce Jillson, UPS 1997 ®} for a LEO states: "Creative energy is limitless, so set aside plenty of time for projects. A romantic involvement with a Libra moves to a deeper level of commitment. Make sure that a relative knows how much you are willing to help with a family project."

    Which part of the "FLaReS" test does the ad fail?

    a. Falsibility
    c. Replicability

    b. Logic
    d. Sufficiency

    It fails all parts of the test:

    a. Falsibility—claims too vague to be evaluated
    b. Logic—not everyone with birthdays July 23 – August 22 have the same circumstances
    c. Replicability—can't reproduce experiment
    d. Sufficiency—no evidence provided

SELF-TEST

Multiple Choice

Select the single best answer for each item. Some items require preliminary calculations before an answer can be selected.

1. Which of the following is not a distinguishing characteristic of science?

    a. invoking supernatural powers
    c. formulation of testable hypotheses
    b. careful measurement
    d. organization of concepts

2. Rachel Carson warned that we might destroy all life on Earth through

    a. overpopulation
    c. chemical warfare
    b. nuclear war
    d. excessive use of pesticides

3. Sir Francis Bacon (1561–1626) predicted that science would endow us with

    a. population problems
    c. new inventions and wealth
    b. terrible weapons of war
    d. pollution problems

4. Which kind of investigator has the least control over the variables in his or her experiment?

    a. biologist
    c. physicist
    b. chemist
    d. sociologist

5. A new pesticide is somewhat effective against insect pests but is also quite toxic to mammals, including humans. The DQ for this pesticide is

    a. very high
    c. somewhat low
    b. somewhat high
    d. uncertain

6. Which has the greatest potential energy?

    a. a small rock moving at high speed at sea level
    b. a large rock moving at low speed at sea level
    c. a large rock balanced at the edge of a mountain top
    d. a large rock falling from the top of a tall building

7. Which of the items mentioned in question 6 has the greatest kinetic energy?

8. An astronaut has a mass of 50.0 kg on Earth. What is her mass on the moon, where gravity is 1/6 that on Earth?

   a. 8.30kg
   c. 83.0kg

   b. 50.0kg
   d. 300. kg

9. An astronaut has a weight of 300 pounds. What is her weight on the moon where the gravity is 1/6 that on Earth?

   a. 8.30 lbs
   c. 83.0 lbs

   b. 50.0 lbs
   d. 300 lbs

10. The physical state that retains volume but not shape is

    a. solid
    c. gas

    b. liquid
    d. none of these

11. A new hair dye is developed that gives good color but is very expensive. The DQ is

    a. very high
    c. somewhat low

    b. somewhat high
    d. uncertain

12. Na is the symbol for

    a. neon
    c. nitrogen

    b. nickel
    d. sodium

13. S is the symbol for

    a. silicon
    c. sodium

    b. silver
    d. sulfur

14. The symbol for potassium is

    a. P
    c. Pm

    b. Po
    d. K

15. The symbol for bromine is

    a. B
    c. K

    b. Bk
    d. Br

16. The symbol for cobalt is

    a. C
    c. CO

    b. Co
    d. CO_2

17. The name of the element with the symbol Ca is

    a. calcium   b. californium
    c. carbon   d. cobalt

18. Which of the following abbreviations stands for a unit of length?

    a. mL   b. mg
    c. dm   d. cc

19. Which of the following abbreviations stands for a unit of mass?

    a. mL   b. mg
    c. dm   d. cc

20. The prefix "micro" is equivalent to

    a. 10^3 b. 10^{-3}
    c. 10^6 d. 10^{-6}

21. Which of the following units of measure is equivalent to cm^3?

    a. mL   b. mm
    c. mg   d. g

22. How long is 1 cm?

    a. 0.01 mm   b. 1 mm
    c. 10 mm   d. 100 mm

23. How many millimeters are there in 10 cm?

    a. 1   b. 10
    c. 100   d. 1000

24. An object that weighs 100 µg also weighs

    a. 0.001 mg   b. 0.01 mg
    c. 0.1 mg   d. 1 mg

25. If a container holds 5 mL, it will hold

    a. 5000 L   b. 0.05 L
    c. $5 cm^3$ d. $0.5 cm^3$

26. What is the density of a sample of kerosene if a 10.0 g sample occupies 13.3 mL?

    a. 0.075 g/mL   b. 0.75 g/mL
    c. 1.33 g/mL   d. 13.3 g/mL

27. If 12 mL of A weighs 4 g, the density of A is

    a. 0.5 g/mL   b. 0.3 g/mL
    c. 2 g/mL   d. 8 g/mL

28. An organic liquid has a density of 0.80 g/cm^3. A sample of 40 cm^3 in volume would have a mass of

    a. 0.020 g   b. 32 g
    c. 50 g   d. 3.2 g

29. What is the volume occupied by 10 g of octane that has a density of 0.70 g/mL?

    a. 7.0 mL   b. 70 mL
    c. 1.4 mL   d. 14 mL

30. On the Kelvin scale, a temperature of 43°C is

    a. 43 K   b. 230 K
    c. 273 K   d. 316 K

31. Liquid nitrogen boils at 77 K. What is this temperature in °C?

    a. −273°C   b. −196°C
    c. 0°C   d. 77°C

32. Research done to lengthen the lifespan of paint is

    a. basic research   b. applied research
    c. technology   d. ecology

33. The kind of science that pursues knowledge for its own sake is called

    a. basic research   b. technology
    c. applied research   d. ecology

True or False

___T___ 34. Chemistry is the study of matter and the changes it undergoes.

___F___ 35. Chemistry is not concerned with changes in energy.

___F___ 36. The United States is a leader among nations using the metric system.

___T___ 37. Manufactured chemical products have greatly affected our lifestyle.

___T___ 38. The ultimate source of nearly all energy on Earth is the sun.

___F___ 39. If measured on the moon, your mass would be different from your mass measured on Earth, although your weight would be the same.

11

ANSWERS

1. a	9. a	17. a	25. c	33. a
2. d	10. b	18. c	26. b	34. T
3. c	11. c	19. b	27. b	35. F
4. d	12. d	20. d	28. b	36. F
5. c	13. d	21. a	29. d	37. T
6. c	14. d	22. c	30. d	38. T
7. d	15. d	23. c	31. b	39. F
8. b	16. b	24. c	32. b	

Atoms

Are They for Real?

KEY TERMS

atom

atomic mass units

atomic theory

law of conservation of mass

law of definite proportions

law of multiple proportions

molecule

periodic table

CHAPTER SUMMARY

2.1 Atoms: The Greek Idea
  - A   The prevailing view of matter held by Greek philosophers in the fifth century B.C. was that of endless divisibility.
  - B.   Democritus, a student of Leucippus, believed that there must be a limit to the divisibility of matter and called his ultimate particles atomos, meaning "indivisible."
  - C.   The Greeks believed that there were only four elements: earth, air, fire, and water.

2.2 Lavoisier: The Law of Conservation of Mass
  - A.   Boyle (1661) proposed that substances capable of being broken down into simpler substances were compounds, not elements.
  - B.   Lavoisier (1780s) helped to establish chemistry as a quantitative science.
    1. The law of conservation of mass states that matter is not created nor destroyed during a chemical change. It is conserved.
       a. We make new materials by changing the way atoms are combined.
    2. Lavoisier was the first to use systematic names for elements.
    3. Lavoisier is often called the "Father of Modern Chemistry."

2.3 Proust: The Law of Definite Proportions
  - A.   Proust (1799) concluded from analyses that elements combine in definite proportions to form compounds and formulated the law of definite proportions (also called the law of constant composition).
  - B.   J.J. Berzelius, Henry Cavendish, William Nicholson, and Anthony Carlisle further proved this law.
  - C.   The law of definite proportions is the basis for chemical formulas.
  - D.   The law of definite proportions also means that compounds have constant properties in addition to constant composition.

2.4 Dalton's Atomic Theory (1803)
- A. Dalton extended the ideas of Lavoisier and Proust with the law of multiple proportions.
  1. Elements may combine in more than one set of proportions with each proportion corresponding to a different compound.
- B. Dalton proposed his atomic theory (model) to explain the laws of chemistry.
  1. A chemical law is a statement that summarizes data obtained from experiments.
  2. A theory is a model that consistently explains observations.
- C. Dalton's Atomic Theory
  1. All matter is composed of small, indestructible and indivisible atoms.
  2. All atoms of a given element are identical, but different elements have different atoms.
  3. Compounds are formed by combining elements in fixed proportions.
  4. A chemical reaction involves a rearrangement of atoms. No atoms are destroyed or broken apart in the process.
- D. Modern Modifications
  1. Atoms can be divided, as we will see in the next chapter.
  2. Atoms can have different masses, as we will see in the next chapter.
  3. Unmodified.
  4. Unmodified for chemical reactions but atoms are broken apart in nuclear reactions.
- E. Explanations using Atomic Theory
  1. Elements are composed of one kind of atom.
  2. Compounds are composed of two or more kinds of atoms chemically combined in definite proportions.
  3. Matter must be atomic to account for the law of definite proportions.
  4. Rearrangement of atoms explains the law of conservation of mass.
  5. The existence of atoms explains how multiple proportions can exist.

2.5 Out of Chaos: The Periodic Table
- A. By the mid-1800s there were 55 known elements, but no successful way had been determined to classify them.
  1. Only relative atomic weights could be determined.
- B. Mendeleev's periodic table (1869) grouped elements by increasing atomic weight.
  1. In some instances, heavier elements were placed before lighter elements in order to group similar properties in the same column.
  2. Blank spots were left in the table for elements that were not yet discovered.

2.6 Atoms: Real and Relevant
- A. "Atom" is a concept that is very useful in explaining chemical behavior.
- B. Atoms are not destroyed in chemical reactions and thus can be recycled.
- C. Materials can be "lost" by scattering their constituent atoms too widely and making it impractical to recover them.

2.7 Leucippus Revisited: Molecules
- A. A molecule is the smallest constituent particle of a compound that still retains the properties of that compound.
- B. Molecules can be divided into atoms.

CHAPTER OBJECTIVES

(You should...)

1. Know that a scientific statement, often mathematical in form, and which summarizes experimental data, is called a <u>law</u>.

2. Know that a substance that can be broken down into two (or more) different simpler substances cannot be an element.

3. Know that chemists create new materials by changing the way atoms combine.

4. Know that Lavoisier gave us the modern names for several elements, including hydrogen and oxygen.

5. Be able to apply the law of definite proportions. (For example, if 3 g of carbon combines with 8 g of oxygen to form 11 g of carbon dioxide, then you would still get only 11 g of CO_2 from 3 g of carbon even if you combined it with 100 g of oxygen).

6. Know that although Democritus and a few others believed in atoms, the prevailing view until about 1800 was that matter was continuous (not atomic).

7. Be able to list the four parts of Dalton's atomic theory. Be able to discuss the modifications to them caused by modern data.

8. Be able to use Dalton's atomic theory to explain elements, compounds, the law of conservation of mass, the law of definite proportions, and the law of multiple proportions.

9. Know that the first attempts at organizing the periodic table grouped the elements by increasing atomic weight and by similar properties.

10. Be able to use the periodic table and atomic number to determine the number of protons and electrons in an atom of any element.

11. Be able to recognize an illustration of the law of multiple proportions.

12. Know that substances with different compositions are different compounds.

DISCUSSION

Chapter 2 is a survey of the history of the atomic theory. The Greek philosopher Democritus, who thought that matter was discontinuous, gave us the word atom. Dalton proposed the first successful attempt to explain the chemical laws of conservation of matter and definite proportion. Dalton's theory also was able to explain the law of multiple proportions. We use the atomic theory because it is useful to explain chemical behavior. You should focus on practicing identifying the interplay between experimenting and theorizing.

EXAMPLE PROBLEMS

1. When 60 g of carbon is burned in air, 220 g of carbon dioxide is formed. How much carbon dioxide is formed when 90 g of carbon is burned?

Using unit conversions (**Appendix C** of the textbook), we can multiply the 90 g of carbon by a conversion factor that preserves the original relationship between 60 g of C and 220 g of CO_2 and then apply it to the current question of increasing the amount of carbon to 90 g.

$$60 \text{ g carbon} = 220 \text{ g carbon dioxide}$$

$$90 \text{ g carbon} = 330 \text{ g carbon dioxide}$$

2. When burned in limited air, 6.0 g of carbon forms 14 g of carbon monoxide. How much carbon monoxide is formed when 360 g of carbon is burned?

$$6.0 \text{ g carbon} = 14 \text{ g carbon monoxide}$$

$$360 \text{ g carbon} = 840 \text{ g carbon monoxide}$$

ADDITIONAL PROBLEMS

1. When 12 g of carbon is burned in air, 44 g of carbon dioxide is formed. How much carbon dioxide is formed when 0.060 g of carbon is burned?

2. When burned in air, 8.0 g of sulfur forms 16 g of sulfur dioxide. How much sulfur dioxide is formed when 400 g of sulfur is burned?

3. Water decomposes, when electricity is passed through it, into hydrogen and oxygen. When 9.0 g of water is electrolyzed, 1.0 g of hydrogen is formed. How much hydrogen is formed when 36 g of water is electrolyzed?

SELF-TEST

Multiple Choice

Select the single best answer.

1. Greek philosophers were not scientists because they

   a. theorized without experimenting
   c. did not vote on their theories

   b. experimented without theory
   d. were not intelligent

2. Democritus's view that matter was atomic in nature was

   a. the accepted view of the ancient Greek philosophers
   b. rejected in Greece but widely accepted in Rome
   c. widely accepted by theologians during the Middle Ages
   d. not widely accepted until the 1800s

3. John Dalton performed a large number of experiments and summarized them all by the statement that two elements can combine in more than one set of proportions. This has been verified during the past century by other scientists and illustrates a scientific

   a. hypothesis
   c. law

   b. theory
   d. model

4. Despite repeated attempts, no one has ever been able to separate chlorine into simpler substances. Therefore we accept chlorine as a(n)

   a. compound
   c. mixture

   b. element
   d. polar molecule

5. "Laughing gas" can be decomposed into two simpler substances, nitrogen and oxygen. Therefore, laughing gas

   a. is an intoxicant
   c. cannot be an element

   b. is a mixture
   d. has the formula NO

6. Natural gas is composed mostly of methane and ethane. Natural gas is

   a. an element
   c. a mixture

   b. a compound
   d. a subatomic particle

7. A scientific statement, often mathematical in form, that summarizes experimental data is called a

   a. law
   c. hypothesis

   b. theory
   d. proposition

8. The fact that 20 g of hydrogen always combines with 160 g of oxygen to form 180 g of water, no matter how much oxygen is present, illustrates the law of

    a. definite proportions   b. multiple proportions
    c. continuity of matter   d. indivisibility of atoms

9. The fact that 20 g of hydrogen combines with 160 g of oxygen to form water and 20 g of hydrogen can also combine with 80 g of oxygen to form hydrogen peroxide, illustrates the law of

    a. definite proportions   b. multiple proportions
    c. continuity of matter   d. indivisibility of atoms

10. All samples of a pure substance

    a. contain no chemicals   b. have the same composition
    c. have a variable composition   d. are heterogeneous

11. Dalton explained the law of conservation of matter by stating that atoms are neither created nor destroyed in a chemical reaction, but they are combined in different ways. Dalton's explanation is an example of a scientific

    a. experiment   b. law
    c. myth   d. theory

12. A scientific law is established through

    a. experimentation   b. legislation
    c. theoretical considerations   d. consulting an expert

13. Water is always composed of 11% hydrogen and 89% oxygen by weight. This statement illustrates the law of

    a. conservation of mass   b. definite proportions
    c. multiple proportions   d. conservation of energy

14. Sulfur forms two oxides in which the weight of oxygen per gram of sulfur is, respectively, 1.00g and 1.50g. This is an illustration of the law of

    a. definite (constant) composition   b. multiple proportions
    c. conservation of matter   d. limiting reagent

15. What limitation exists on the recycling of copper? We must

    a. be careful not to destroy copper atoms
    b. find a way to melt copper
    c. not scatter copper atoms too widely
    d. find a way to change copper atoms into other kinds of atoms

16. The results of many experiments are summarized in the statement that the force of repulsion between two like electrical charges is proportional to the distance between the charges. This is an example of a scientific

    a. hypothesis   b. law
    c. myth   d. theory

17. The idea that matter is continuous rather than atomic prevailed for 2000 years because the intellectual leaders of the time failed to test their ideas by making

    a. hypotheses   b. logical deductions

    c. experiments   d. theories

18. Which set of compounds illustrates the law of multiple proportions?

    a. CH_4, CO, CCl_4 b. N_2O, NO, NO_2

    c. NaCl, NaBr, NaI   d. all of these

19. According to Dalton's atomic theory, atoms

    a. are components of all matter   b. are indivisible

    c. combine in small whole-number ratios   d. all of these

20. In forming new materials, chemists usually

    a. create new kinds of atoms   b. split atoms into parts

    c. change the way atoms are combined   d. change the weight of atoms

21. Elements are placed in groups in the periodic table on the basis of

    a. alphabetical listing   b. similar chemical properties

    c. number of neutrons in the nucleus   d. order of discovery

22. Mendeleev organized his periodic table of elements by

    a. alphabetical listing of names   b. increasing atomic number

    c. date of discovery   d. physical state

True or False

___ 23. The prevailing view of the ancient Greek philosophers was that matter is continuous, not atomic.

___ 24. Dalton is known as the "Father of Modern Chemistry."

___ 25. Dalton regarded the atom as hard and indivisible.

___ 26. If lead sulfide were made on the moon, it would differ in composition from that made on Earth.

___ 27. Carbon reacts with oxygen to form only one compound.

___ 28. The Roman poet Lucretius presented a strong argument for the atomic nature of matter in his long poem, *On the Nature of Things*.

___ 29. The law of definite proportions is consistent with the theory of the continuity of matter.

___ 30. Two elements may combine in more than one set of proportions.

___ 31. According to the information in Figure 2.6 in your text, ten grams of lead and 10 grams of sulfur yield 20 grams of lead sulfide.

_____ 32. According to the information in Figure 2.6 in your text, 10 atoms of lead and 10 atoms of sulfur yield 10 molecules of lead sulfide.

Matching

Note: A name may be associated with more than one concept.

Concept	*Name of Person*
C 33. Matter is indivisible	a. Boyle
e 34. "Father of Chemistry"	b. Proust
a 35. Compounds could be broken down into simpler substances	c. Democritus
e 36. First to use systematic names	d. Dalton
b 37. Proposed definite proportions	e. Lavoisier
d 38. First successful atomic model based on experimental data	

ANSWERS

Additional Problems

1. 0.22 g carbon dioxide
2. 800 g sulfur dioxide
3. 4.0 g hydrogen

Self-Test

1. a	9. b	17. c	25. T	33. c
2. d	10. b	18. b	26. F	34. e
3. c	11. d	19. d	27. F	35. a
4. b	12. a	20. c	28. T	36. e
5. c	13. b	21. b	29. F	37. b
6. c	14. b	22. b	30. T	38. d
7. a	15. c	23. T	31. F	
8. a	16. b	24. F	32. T	

Atomic Structure

Images of the Invisible

KEY TERMS

alkali metals	excited state	nucleon number
alkaline earth metals	gamma rays	nucleus
alpha particle	ground state	orbital
anion	groups	periods
atomic number	halogens	photon
beta particles	ions	proton
cathode ray	isotopes	quantum
cation	main group elements	radioactivity
deuterium	mass number	sublevel
electrodes	metalloids	transition elements
electrolysis	metals	tritium
electrolyte	neutron	valence electrons
electron	noble gases	X-rays
electron configuration	nonmetals	
energy levels	nucleon	

CHAPTER SUMMARY

3.1 Electricity and the Atom
    A. Volta invented an electrochemical cell (1800) called the "voltaic pile."
    B. Electrolysis
        1. Humphrey Davy produced elements from compounds by passing electricity through them (electrolysis).
        2. Faraday continued to work.
        3. Electrolytes are compounds that conduct electricity when melted or taken into solution.
        4. Electrodes are carbon rods or metal strips that carry electric current when inserted into a molten compound or solution.
        5. The anode is the positively charged electrode. The cathode is the negatively charged electrode.

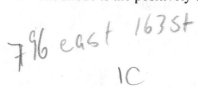

6. The anion is a negatively charged ion attracted to the anode. A cation is a positively charged ion attracted to the cathode.
C. Crookes (1875) discovered cathode rays by passing electricity through a partially evacuated gas discharge tube.
   1. Cathode rays travel from the cathode to the anode.
D. Thomson (1897) found that cathode rays were deflected in an electric field and thus must contain charged particles.
   1. These negatively charged particles were called electrons and were found to be the same for all gases used to produce them.
   2. Cathode rays travel in straight lines in the absence of an applied field.
   3. Thomson calculated the ration of the electron's mass to its charge.
E. Goldstein (1886) used an apparatus similar to Crookes's tube to study positive atomic particles.
   1. The positively charged particles were found to be more massive than electrons and to vary with the type of gas used in the experiment.
   2. The lightest positive particle obtained was derived from hydrogen and had a mass 1837 times heavier than that of electrons.
F. Millikan (1909) determined the charge of the electron.
   1. The charge on an electron is –1.
   2. The mass of the electron is 9.110×10^{-28} g.

3.2 Serendipity in Science: X-Rays and Radioactivity
A. Roentgen (1895) discovered X-rays.
B. Becquerel discovered radioactivity while studying fluorescence.
C. Marie and Pierre Curie studied radioactivity.
   1. They discovered the elements radium and polonium.

3.3 Three Types of Radioactivity
A. Rutherford classified three types of radioactivity.

Name	Symbol	Mass	Charge
1. Alpha rays	α particles	4 amu	2+
2. Beta rays	β particles	1/1837 amu	1-
3. Gamma rays	γ rays	0	0

3.4 Rutherford's Experiment: The Nuclear Model of the Atom
A. Rutherford's experiment showed that the positive charge and nearly all the mass are concentrated in the core (nucleus).
B. Rutherford (1914) proposed that protons constitute the positively charged matter of all atoms, not just that of hydrogen.

3.5 Structure of the Nucleus
A. Chadwick (1932) discovered a nuclear particle, called a neutron, which has about the same mass as a proton, but has no charge.
   1. The number of protons in the nucleus (atomic number) determines the identity of the atom.
B. Isotopes have the same number of protons (that is, the same atomic number) but a different number of neutrons (different atomic mass).

3.6 Electron Arrangement: The Bohr Model
A. Flame tests rely on the color of flames to identify elements.
   1. A prism separates the light of the flame into specific colored lines.
   2. These different colored lines represent different wavelengths or energies.

B. A line spectrum is a pattern of lines, each line corresponding to different wavelengths or energies emitted by an element. A continuous spectrum contains all colors of the spectrum (and all wavelengths and all energies).
   1. Some lines are in the infrared or ultraviolet regions and cannot be seen by the unaided eye.
C. Line spectra can be generated by emission or absorption processes.
   1. Emission spectra-light given off by elements when they are heated.
   2. Absorption spectra-light left after elements have absorbed certain frequencies of light.
D. Bohr's explanation of line spectra stated that discrete spectra arise because the energy of the electrons in an atom is quantized, meaning it can only absorb discrete values of energy.
   1. Energy levels: specified energy values for an atom.
   2. Ground state: electrons in the lowest possible energy levels.
   3. Excited state: due to added energy, an electron jumps to a higher energy level.
   4. The maximum number of electrons in any given energy level is given by the formula $2n^2$.
E. Ground State and Excited State
   1. Ground state: All electrons are in lowest possible energy levels (this is stable).
   2. Excited state: One (or more electrons) is elevated to a higher energy level leaving an empty lower energy level open below it (this is unstable).
F. Building atoms
   1. Add electrons to energy levels closest to nucleus.
   2. Once an excited energy level is filled, the electrons go into the next higher level.

3.7. The Quantum Mechanical Atom
A. de Broglie suggested the wavelike properties of electrons.
B. Schrödinger developed equations to describe the behavior of electrons in atoms.
C. Electrons move in specifically shaped volumes of space.
   1. These different shapes are indicated by the letters s, p, d, and f.
   2. The s sublevel has one orbital.
       p sublevel has three orbitals.
       d sublevel has five orbitals.
       f sublevel has seven orbitals.
   3. An orbital can have a maximum of two electrons.
D. Electron configurations are expressed by numbers that indicate the main energy level and letters that indicate the sublevel. For example, the notation $2p^5$ indicates that there are 5 electrons in the p orbitals of the second energy level.

3.8. Electron Configurations and the Periodic Table
A. The modern periodic table is arranged in order of increasing atomic number and grouped according to electronic structure.
B. Groups (families)—the vertical columns.
   1. Elements in one group have similar chemical properties.
     a. The groups (families have similar outer electron configurations.
   2. Groups are designated by a Roman numeral and letter "A" or "B".
     a. New IUPAC system renumbers groups 1 to 18 and does not use "A" or "B" designations.
   3. Group A elements—main group elements – Members of the same A group have the same number of electrons in the outermost shell.
     a. Family Groups
        1. Group IA—Alkali Metals (highly reactive) ns^1
        2. Group IIA—Alkaline Earth Metals (moderately reactive) ns^2
        3. Group VIIA—Halogens (highly reactive nonmetals) ns^2np^5
        4. Group VIII—Noble Gases (non-reactive nonmetals) ns^2np^6
   4. Group B elements—transition group elements

a. These elements have properties that are "transition" as there electron configurations are more complex.

C. Metals, nonmetals, and metalloids.
   1. Elements in the periodic table are divided into two main classes: metals and nonmetals. A heavy, stair-like line separates the two classes.
      a. Metals—elements to the left of the line
      b. Nonmetals—elements to the right of the line
      c. Metalloid—elements bordering the line.
   2. Characteristics of metals: luster, good conductors of hear and electricity, solid at room temperature (except mercury, a liquid), malleable, and ductile.
   3. Characteristics of nonmetals: lack metallic properties.
      a. Most are solids or gases at room temperature.
      b. Bromine is a liquid.
   4. Characteristics of metalloids: properties of both metals and nonmetals.
   5. The group number for the main group elements (A groups) gives the number of electrons in the outer energy level (valence electrons).

E. Periods
   1. The period tells us how many main electron energy levels an atom has.

3.9. Which Model to Choose?
   A. There are several different models of atoms.
   B. We use the model that is most helpful in understanding a particular concept.

CHAPTER OBJECTIVES

(You should...)

1. Be able to identify and locate (in or outside the nucleus) the fundamental parts of an atom.

2. Be able to compare and contrast the characteristics of alpha, beta, and gamma radiation.

3. Be able to match the scientists mentioned in this chapter (and in Chapter 2) with their contributions to our concept of the atom and its structure.

4. Be able to draw Bohr diagrams for the first 20 elements.

5. Be able to define or identify each of the Key Terms.

6. Know that cathode rays are streams of electrons.

7. Know that line spectra are used to identify elements.

8. Be able to explain line spectra in terms of electron structure.

9. Be able to differentiate between emission and absorption spectra.

10. Be able to give the charge and approximate mass of the proton, the neutron, and the electron.

11. Be able to describe the location of the parts of an atom.

12. Know that for the first 20 elements, the number of electrons in the outer shell is equal to the group number. (ex. chlorine is in Group VIIA; a chlorine atom has 7 electrons in its outer shell)

13. Be able to determine the number of energy levels in an atom from its period and the periodic table.

14. Be able to select elements that are in the same group on the periodic table.

15. Be able to select elements that are in the same period on the periodic table.

16. Be able to tell whether an element is a metal, nonmetal, or a metalloid from its position on the periodic table.

17. Be able to identify groups of elements in the periodic table that are classified as alkali metals, alkaline earth metals, halogens, noble gases, and transition metals.

18. Be able to determine the electron configuration of the first 20 elements.

DISCUSSION

We can see the images of atoms even though they are too small to be seen by visible light. The experiments that led to the modern atomic theory begin with the beginnings of electricity and radioactivity and go on through spectroscopy. The descriptions of the arrangement of electrons around the nucleus and quantum mechanics are the result of many years of the interplay between experimentation and the proposal of hypotheses. Finally, we organize the periodic table to maximize our understanding of atomic structure.

SELF-TEST

Multiple Choice

You may refer to the periodic table (**see inside front cover**).

1. Two subatomic particles found in the nucleus of an atom are

    a. neutrons and electrons
    c. protons and neutrons

    b. electrons and protons
    d. beta particles and protons

2. The nucleus contains

    a. all of the negative charge, none of the mass
    b. all of the positive charge, none of the mass
    c. all of the negative charge, all of the mass
    d. all of the positive charge, all of the mass

3. An alpha particle is the same as

    a. its nucleus
    c. lithium nucleus

    b. helium nucleus
    d. carbon nucleus

25

4. A beta particle is the same as

    a. an electron   b. a proton
    c. a neutron   d. a photon

5. Electricity interacts with matter because

    a. atoms contain charged particles   b. atoms have mass
    c. atoms are quantized   d. atoms have orbitals

6. Three types of radiation emitted during radioactive decay are

    a. alpha, beta, gamma   b. electron, proton, neutron
    c. positron, neutron, proton   d. electron, proton, X-ray

7. A subatomic particle with a mass of 1 amu and a charge of 1+ is the

    a. electron   b. neutron
    c. positron   d. proton

8. Rutherford, through his gold foil experiment, discovered that the atomic nucleus is

    a. tiny and not very dense   b. large but not very dense
    c. large and very dense   d. tiny but very dense

9. Which element commonly has no neutrons in its nucleus?

    a. H   b. He
    c. C   d. F

10. The arrangement of electrons around the nucleus is quantized is proven by

    a. continuous spectra   b. line spectra
    c. infrared spectra   d. X-ray spectra

11. Cathode rays are

    a. alpha particles   b. electrons
    c. gamma rays   d. protons

12. How many electrons are there in a neutral atom of S?

    a. 11   b. 14
    c. 16   d. 21

13. How many electrons are there in a neutral atom of Na?

    a. 11   b. 12
    c. 22   d. 23

14. How many protons are there in an atom of Be?

    a. 4   b. 5
    c. 9   d. 11

15. How many protons are there in Ca?

    a. 6   b. 17

    c. 20   d. 27

16. What is the symbol of the element with atomic number 51?

    a. Cr   b. Sb

    c. V   d. Zr

17. Which element is in the same group as Si?

    a. S   b. As

    c. Sn   d. N

18. Si is classified as a

    a. metal   b. nonmetal

    c. metalloid   d. none of these

19. Which element is in the same period as Al?

    a. As   b. B

    c. C   d. Mg

20. The number of protons in the nucleus of an Mg atom is

    a. 12   b. 24

    c. 36   d. variable

21. Mg is classified as a

    a. metal   b. nonmetal

    c. metalloid   d. none of these

22. Which element is an alkali metal?

    a. Na   b. Be

    c. Fe   d. Al

23. What is the symbol of the element with the atomic number 28?

    a. N   b. Si

    c. Ni   d. Ba

24. Which element is a noble gas?

    a. H   b. He

    c. O   d. F

25. Which element is a halogen?

    a. H   b. He

    c. O   d. F

26. Which of the following elements is a transition metal?

    a. Ti   b. Sr
    c. Te   d. Be

Use the Bohr model to answer questions 27-36.

27. The electron configuration of calcium is

    a. 2, 4   b. 2, 8, 10
    c. 2, 8, 8, 2   d. 2, 8, 2, 8

28. The electron configuration of phosphorus is

    a. 2, 8, 5   b. 3, 8, 4
    c. 2, 3, 10   d. 2, 8, 19

29. The electron configuration of chlorine is

    a. 2, 8, 15   b. 2, 8, 7
    c. 2, 7, 8   d. 17, 18

30. The electron configuration of potassium is

    a. 2, 8, 8, 1   b. 2, 8, 9
    c. 2, 8, 18, 9   d. 2, 8, 18, 8, 1

31. How many electrons are there in the outer energy level of a nitrogen atom?

    a. 0   b. 3
    c. 5   d. 7

32. All these atoms have seven electrons in the outermost energy level except

    a. H   b. F
    c. Cl   d. Br

33. How many electrons are there in the outer energy level of a sulfur atom?

    a. 2   b. 4
    c. 6   d. 16

34. How many electrons are in the outermost energy level of Group VIIA?

    a. 1   b. 3
    c. 5   d. 7

35. How many electrons are in the outermost energy level of Group IIA elements?

    a. 1   b. 2
    c. 6   d. 11

36. What is the maximum number of electrons in the second energy level?

    a. 4   b. 8

    c. 10   d. 18

Use the quantum mechanical model to answer questions 37-43

37. How many electrons exist in the same orbital?

    a. 1   b. 2

    c. 6   d. 8

38. The electron configuration of Be is

    a. $1s^2 2s^2$ b. $1s^2 2s^2 2p^5$

    c. $1s^2 2p^2$ d. $1s^2 2p^7$

39. What is the maximum number of electrons that can be placed in the 2p orbitals of an atom

    a. 2   b. 4

    c. 6   d. 12

40. The electron configuration of oxygen is

    a. $1s^2 2p^4$ b. $1s^2 2s^2 2p^4$

    c. $1s^2 2p^4 3d^2$ d. none of these

41. What element has the electron configuration $1s^2 2s^2 2p^5$?

    a. B   b. N

    c. F   d. none

42. What is the atomic number of the element with the electron configuration $1s^2 2s^2 p^3$?

    a. 3   b. 5

    c. 7   d. none of these

43. What is wrong with the electron configuration $1s^2 1p^6 2s^2 3p^1$?

    a. too many electrons in 1s   b. there is no 1p orbital

    c. the 2s orbital can hold 3 electrons   d. the 2p orbital is not filled

44. The diagram shows the

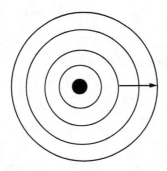

   a. excitation from n = 1 to n = 4
   b. excitation from n = 2 to n = 4
   c. the relaxation of an electron from n = 4 to n = 1
   d. the relaxation of an electron from n = 4 to n = 2

45. Which of the following describes a metal?

   a. shiny, ductile, malleable, good conductor
   b. dull, ductile, good conductor
   c. shiny, brittle, good conductor
   d. shiny, ductile, non-conductor

46. Quantum mechanics describes the electron as

   a. probability shapes of electrons
   b. having only certain energies
   c. having wave-like properties
   d. all of the above

Matching

Note: A name may be associated with more than one concept.

Concept	*Name of Person*
47. proponent of the atomic structure of matter among the ancient Greeks	a. Becquerel
48. proposed the modern atomic theory	b. Bohr
49. law of conservation of mass	c. Chadwick
50. law of constant composition	d. Curie
51. law of definite proportions	e. Dalton
52. law of multiple proportions	f. Democritus

_g_ 53. made quantitative measurements the standard of chemical experimentation

_k_ 54. the nuclear atom (mass concentrated in the nucleus)

_b_ 55. the atom as a miniature solar system

_l_ 56. described atomic structure with complex mathematical equations

_m_ 57. discovered the electron

_c_ 58. discovered the neutron

_b_ 59. explained the line spectra of the elements

_j_ 60. discovered X-rays

_a_ 61. discovered radioactivity

_d_ 62. discovered new radioactive elements

g.   Lavoisier

h.   Einstein

i.   Proust

j.   Roentgen

k.   Rutherford

l.   Schrödinger

m.   Thomson

ANSWERS

1. c	12. c	23. c	34. d	44. b	54. k
2. d	13. a	24. b	35. b	45. a	55. b
3. b	14. a	25. d	36. b	46. d	56. l
4. a	15. c	26. a	37. b	47. f	57. m
5. a	16. b	27. c	38. a	48. e	58. c
6. a	17. c	28. a	39. c	49. g	59. b
7. d	18. c	29. b	40. b	50. i	60. j
8. d	19. d	30. a	41. c	51. i	61. a
9. a	20. a	31. c	42. c	52. e	62. d
10. b	21. a	32. a	43. b	53. g	
11. b	22. a	33. c			

32

Nuclear Chemistry

The Heart of the Matter

KEY TERMS

background radiation	half-life	photoscan
binding energy	hydrogen bomb	positron
C-14 dating	induced radiation	radioactive decay
chain reaction	ionizing radiation	radioisotopes
cosmic rays	mass-energy equation	thermonuclear reactions
critical mass	nuclear fission	tracers
electron capture	nuclear fusion	transmutation
fundamental particles	nuclear winter	

CHAPTER SUMMARY

4.1   Natural Radioactivity: Nuclear Equations
   A.   An oversimplified but useful chemical view of the atomic nucleus is that it consists of protons, neutrons, and the force that holds them together.
      1.   Proton   (p+)   1.0073 amu   charge 1+
      2.   Neutron   (n)   1.0087 amu   charge 0
      3.   Electron   (e–)   0.0005 amu   charge 1–
   B.   The atomic number (Z) of an element is determined by the number of its protons.
   C.   Atoms with the same number of protons but a different number of neutrons are called isotopes.
      1.   All isotopes of a given element have similar chemical properties.
      2.   Water made from the deuterium isotopes of hydrogen is called "heavy water."
   D.   Positron emission and electron capture are two more types of nuclear reactions.
      1.   Positron example

$$^{11}_{6}C \longrightarrow ^{0}_{+1}e + ^{11}_{5}B$$

      2.   Electron capture example

$$^{195}_{79}Au + ^{0}_{-1}e \longrightarrow ^{195}_{78}Pt$$

   E.   Some Differences between Chemical and Nuclear reactions
      1.   Atoms in chemical reactions retain their identity but may change in nuclear reactions.
      2.   Chemical reactions involve electrons while nuclear reactions mainly involve protons and neutrons.

3. Chemical reaction rates can be changed with changes in temperature while nuclear reaction rates are not affected by changes in temperature.
4. The energy of chemical reactions is very small when compared to nuclear reactions
5. Mass is conserved in chemical reactions while mass and energy is conserved in nuclear reactions.

F. Nuclear Arithmetic: Symbols for Isotopes
1. Protons and neutrons are collectively referred to as nucleons.

   X = chemical symbol

   $$^{A}_{Z}X$$ where A = nucleon number (sum of p + n)
   Z = atomic number (number of p)

   a. Number of neutrons = A − Z
   b. The nucleon number of an isotope is written as a suffix to the name.
   c. Hydrogen-2 is the symbol for deuterium
2. The atomic weight listed in the periodic table is the isotopically weighted average value for that element.

G. Some nuclei are unstable and undergo spontaneous reactions called radioactive decay.
H. There are three major types of radioactivity.
1. Alpha decay is characterized by the giving off of particles identical to a helium nucleus and consisting

   of two protons plus two neutrons $\left(^{4}_{2}He\right)$

1. A nuclear change whereby one element is changed into another is called <u>transmutation</u>.

   $$^{226}_{88}Ra \longrightarrow {}^{222}_{86}Rn + {}^{4}_{2}He$$

2. Beta decay is characterized by the emission of a nuclear electron $\left(^{0}_{-1}e\right)$ as neutron is

   $$^{14}_{6}C \longrightarrow {}^{14}_{7}N + {}^{0}_{-1}e$$

3. Gamma decay is characterized by the emission of energy (not particles).

4.2. Half-life
A. Radioactivity is dependent on the isotope involved, but is generally independent of any outside influences such as temperature or pressure.
B. Radioactivity is a random process. Large numbers of atoms have a predictable half-life characteristic of that isotope.
1. The half-life is that period of time during which one half of the radioactive atoms undergo decay.
   a. The fraction of the original radioactive sample remaining after n half-lives is given by $(1/2^{n})$, or 1/2, 1/4, 1/8 1/16, and so forth.

4.3. Radioisotope Dating
A. The half-lives of radioisotopes can be used to estimate the age of rocks and artifacts.
B. Shroud of Turin dated between a.d. 1260 and 1340. Dead Sea Scrolls dated 2000 years ago.
C. Many artifacts are dated by their carbon-14 content.
D. Tritium dating is used to date items up to 100 years old.

4.4. Artificial Transmutation
   A. Nuclear changes can also be brought about by the bombardment of stable nuclei with subatomic particles such as, β, or neutron

$$\,^{9}_{4}\text{Be} \;+\; \,^{4}_{2}\text{He} \;\longrightarrow\; \,^{12}_{6}\text{C} \;+\; \,^{1}_{0}\text{n}$$

      1. A nuclear change whereby one element is changed into another is called transmutation.
   B. Transmutations carried out by Rutherford and later by Chadwick with alpha particles led to the discovery of fundamental particles such as the neutron.

4.5 Induced Radioactivity
   A. Iréne Curie and her husband Frédéric Joliot won the Nobel prize for creating a radioactive isotope of phosphorus that emitted a positron $\left(\,^{0}_{+1}\text{e}\right)$.
      1. A positron is equal in mass to an electron but opposite in charge.

4.6. Uses of Radioisotopes
   A. Radioactive isotopes can substitute chemically for their non-radioactive counterparts, but have the advantage of being easily detected; they act as tracers.
      1. Radiographs are photographic records of the movement of a tracer in an object.
   B. Radioisotopes are used to kill microorganisms in the irradiation of food.

4.7. Nuclear Medicine
   A. Radiation therapy: Radiation from radioisotopes is used to treat cancer.
      1. Radiation is more lethal to the rapidly reproducing cancer cells than to normal cells.
   B. Radioisotopes are used in medical diagnosis.
      1. Iodine-131—size, shape, and activity of the thyroid gland.
      2. Gadolinium-153—bone mineralization, osteoporosis
      3. Technetium-99m (m = metastable)—used for diagnostic tests. It emits gamma radiation and then reverts to a stable isotope.
   C. Computer-aided medical imaging methods such as positron emission tomography (PET) scans use positron-emitters to generate gamma rays inside the body and measure dynamic processes.

4.8 Radiation and Us
   A. Ionizing radiation: Radiation with enough energy to knock electrons from atoms and molecules converting them to ions
   1. Ionizing radiation interferes with normal chemical processes of cells.
   B. Background radiation: Radiation from natural radioactive isotopes found in cosmic rays, air, water, soil and rocks
   C. The second leading source of our exposure to radiation is medical X-rays.
   D. Radiation damage to cells
      1. Radiation causes changes in DNA that produce mutations in offspring.
   E. Penetrating Power of Radiation
      1. The penetrating power of different types of radiation varies widely with gamma>beta>alpha. This is due in part to their masses but also to their charges.
      2. Alpha particles outside the body do little damage because they can't penetrate the skin. Inside the body, alpha particles inflict great damage because they are trapped in a small area.
      3. Protection from radiation.
         a. Move away from the source.
         b. Use shielding between you and the source of the radiation.

4.9. Einstein and the Equivalence of Mass and Energy
   A. Albert Einstein (1905) derived a relationship between matter and energy as a part of his theory of relativity.
   B. One gram of matter is the energy equivalent of heating a home for 1,000 years.

$$E = mc^2 \qquad \text{where} \qquad \begin{aligned} E &= \text{energy} \\ m &= \text{mass} \\ c &= \text{speed of light} \end{aligned}$$

   C. Binding energy is equivalent to the difference in mass between the individual protons and neutrons and the mass of the nucleus they form; related by the equation $E = mc^2$.
   D. Elements with the highest binding energy are the most stable.
   E. Nuclear fusion: a reaction that combines small atoms and releases great amounts of energy.

4.10 The Building of the Bomb
   A. Nuclear fission refers to the process of splitting heavy atomic nuclei into major fragments. Enormous amounts of energy are released during this process.
   B. Nuclear chain reaction—neutrons released in fission of one atom trigger fission in other atoms setting off a chain reaction.
   C. Manhattan project history.
   D. For use in bombs, uranium-238 must be enriched to 90% uranium-235.
   E. Synthesis of plutonium—neutron bombardment of U-238 leads to the fissionable isotope, plutonium-239.
   F. The critical mass is the minimum amount of fissionable material needed to sustain a chain reaction.
   G. Research to build the bomb was conducted at a secret lab in Los Alamos, NM and tested on July 16, 1944.
   H. During World War II, the United States dropped two nuclear bombs on Japan.
      1. August 6, 1945: A uranium bomb was dropped on Hiroshima.
      2. August 9, 1945: A plutonium bomb was dropped on Nagasaki.
      3. August 14, 1945: Japan surrendered.

4.11 Radioactive Fallout
   A. Nuclear fission results in the formation of radioactive daughter isotopes. Several of these daughter isotopes are dangerous when they are incorporated into the food chain.
      1. Strontium-90 can substitute for calcium and is incorporated into the bone.
      2. Iodine-131 is concentrated in the thyroid.
      3. Cesium-137 mimics potassium.

4.12 Nuclear Winter
   A. Nuclear explosions produce enormous amounts of dust, soot, and smoke.
      1. This dust, soot, and smoke could block out the sun. The Earth would be cold and dark; a "nuclear winter" would set in.
4.13 Nuclear Power Plants
   A. Nuclear energy accounts for one-fifth of the energy produced in the United States.
   B. Heat from nuclear reactions is used to produce steam to turn turbines for generating electricity.
   C. Nuclear energy is cleaner than fossil fuel energy, but nuclear energy produces a lot of waste heat and radioactive waste.

4.14 Thermonuclear Reactions
   A. Thermonuclear reactions require high temperatures for initiation.
      1. In stars like our sun, thermonuclear reactions cause light nuclei like hydrogen to fuse, forming helium and releasing enormous amounts of energy.
   B. Hydrogen bombs utilize a small fission bomb to create the high temperatures needed for the fusion process.

36

4.15 The Nuclear Age
    A. New radioactive elements have been used to save lives through diagnostic and therapeutic techniques.

CHAPTER OBJECTIVES

(You should...)

1.   Be able to calculate the number of neutrons, the number of protons, or the nucleon number given any two of these.

2.   Be able to supply the missing nucleide in any nuclear reaction.

3.   Be able to list the differences between chemical and nuclear reactions.

4.   Know that beta particles are identical to high-speed electrons.

5.   Know that alpha particles are identical to helium nuclei.

6.   Know the symbolism for alpha, beta, and gamma radiation.

7.   Be able to recognize nucleides that are isotopes (they have the same atomic number).

$$\alpha = {}^{4}_{2}He \qquad \beta = {}^{0}_{-1}e \qquad \gamma = {}^{0}_{-1}\gamma$$

8.   Recognize that gamma rays are not streams of particles; they are electromagnetic radiation similar to X-rays.

9.   Be able to give the mass (in amu) and charge (in electronic charge units) of protons, neutrons, electrons, alpha particles, beta particles, and gamma rays.

10.   Be able to distinguish between nuclear fission and nuclear fusion and tell how each results in the release of energy.

11.   Know that ${}^{235}_{92}U$ and ${}^{239}_{93}Pu$ are fissionable.

12.   Know that nuclear power plants produce about one-fifth of the energy in United States.

13.   Know that the sun is a nuclear fusion reactor.

14.   Know that nuclear fusion uses the heavy isotopes of hydrogen $\left({}^{2}_{1}H, \text{ called deuterium , and } {}^{3}_{1}H, \text{ called tritium} \right)$ as fuel.

15.   Be able to calculate the age of an artifact, given the fractional decrease (1/2, 1/4, 1/8, etc.) in activity and the half-life of the isotope undergoing radioactive decay.

16.   Know that our principal exposure to radiation is from natural sources (background radiation). The second greatest source is medical X rays.

17.   Know that gamma rays are the most penetrating and alpha particles the least penetrating radiations.

18. Know that once inside the body, alpha particles are the most damaging of the three types of radiation.

19. Be able to describe several uses of radioisotopes.

20. Be able to list several radioisotopes and describe their use in medicine.

21. Be able to describe how a PET scan works.

22. Be able to explain what is meant by "nuclear winter."

23. Be able to define or explain each of the Key Terms for this chapter.

DISCUSSION

The nucleus is 100,000 times as small as the atom yet it contains all the mass and all the positive charge. Nuclear symbols summarize the information needed to describe an isotope: the number of protons, (atomic number or charge) and the number of neutrons (atomic mass number). In a nuclear equation, the atomic numbers and mass numbers are conserved, that is, equal before and after the reaction. The half-life of each isotope is characteristic of that isotope and a constant that cannot be changed. (A half-life is the time for one-half of the material to decay.) We can cause transmutations by bombarding nuclei with subatomic particles. We make use of radioisotopes according to their penetrating power and the amount of change they can cause in medical, food, and industrial applications. The age of rocks and archeological artifacts are determined using half-lives. In a nuclear reaction, mass is converted into energy producing a million times the energy of an equivalent chemical reaction. This energy is used in fission (splitting nuclei) processes such as the A-bomb and current nuclear power plants. These processes require sophisticated technology to increase the amount of fissionable material ("enrich") and have hazards associated with radioactivity. Thermonuclear (fusion—adding nuclei together) is the reaction of the sun and H-bomb.

EXAMPLE PROBLEMS

1. Thorium-232 $\left({}^{232}_{90}\text{Th} \right)$ undergoes alpha decay. What new element is formed?

   Mass and charge are conserved. Alpha particles are helium nuclei and thus carry away 4 mass units and 2 units of charge. The new nuclei must have a mass of $232-4 = 228$ and a nuclear charge of $90 - 2 = 88$. The nuclear charge (atomic number) identifies the new element as radium. The equation for the process is

$$ {}^{232}_{90}\text{Th} \longrightarrow {}^{4}_{2}\text{He} + {}^{238}_{88}\text{Ra} $$

2. Iodine-131 $\left(\begin{array}{c} 131 \\ 53 \end{array} I \right)$ undergoes beta decay. What new element is formed?

The beta particle takes away essentially no mass (the nucleon number remains unchanged) and a charge of 1–. Subtracting 1– increases the nuclear charge by one. The new element has a nuclear charge of 54, identifying it as xenon. The equation is

$$\begin{array}{c} 131 \\ 53 \end{array} I \longrightarrow \begin{array}{c} 0 \\ -1 \end{array} e + \begin{array}{c} 131 \\ 54 \end{array} Xe$$

3. Radioactive nitrogen-13 has a half-life of 10 minutes. After an hour, how much of this isotope would remain in a sample that originally contained 96 mg?

One hour is 60 minutes or 6 half-lives (n = 6) and m = 96 mg.

$$m_r = \left(\frac{1}{2^6} \right) \times (96 \text{ mg}) = \left(\frac{1}{64} \right) \times (96 \text{ mg}) = 1.5 \text{ mg}$$

4. Radioactive thalium-154 has a half-life of 5 seconds. After 10 seconds, how many milligrams of this isotope remain in a sample that originally contained 160 mg?

Ten seconds is 2 half-lives (n = 2), and m0 =160 mg.

$$m_r = \left(\frac{1}{2^2} \right) \times (160 \text{ mg}) = \left(\frac{1}{4} \right) \times (160 \text{ mg}) = 40 \text{ mg}$$

ADDITIONAL PROBLEMS

1. Complete the following equations by supplying the missing component.

a. $$\begin{array}{c} 234 \\ 90 \end{array} Th \longrightarrow \begin{array}{c} 0 \\ -1 \end{array} e + ? \overset{234}{\underset{89}{}}$$

b. $$\begin{array}{c} 222 \\ 86 \end{array} Rn \longrightarrow \begin{array}{c} 4 \\ 2 \end{array} He + ? \overset{226}{\underset{80}{}}$$

c. $$\begin{array}{c} 56 \\ 26 \end{array} Fe + \begin{array}{c} 2 \\ 1 \end{array} H \longrightarrow \begin{array}{c} 54 \\ 25 \end{array} Mn + ?$$

39

2. Protactinium-234 has a half-life of 1 minute. How much of a 400-µg sample of protactinium would remain after 1 minute? After 2 minutes? After 4 minutes?

3. The half-life of plutonium-239 is 24,300 years. About 8 kg of this isotope is released in a nuclear explosion. How many years would pass before the amount was reduced to 1 kg?

SELF-TEST

Multiple Choice

1. The nucleus

    a. contains all of the mass
    b. contains all of the positive charge
    c. is ten thousand times as small as the rest of the atom
    d. all of the above

2. How many protons does potassium have?

    a. 1   b. 19
    c. 38   d. 76

3. Which isotope has the largest number of neutrons in the nucleus?

    a. $^{120}_{50}Sn$ b. $^{123}_{51}Sb$

    c. $^{130}_{52}Te$ d. $^{127}_{53}I$

4. How many neutrons are there in a $^{169}_{70}Yb$ nucleus?

    a. 70   b. 99
    c. 169   d. 239

5. Isotopes have the same

    a. nucleon number   b. atomic number
    c. atomic mass   d. half-life

6. Isotopes have the same number of

    a. protons   b. neutrons
    c. photons   d. beta particles

7. Which is an isotope of potassium–39?

a. $^{40}_{19}K$

b. $^{39}_{18}Ar$

c. $^{39}_{20}Ca$

d. $^{89}_{39}Y$

8. Which subatomic particle has a mass of approximately 1 amu and a charge of zero?

a. proton
c. electron

b. neutron
d. gamma ray

9. Alpha particles are identical to

a. helium (He) nuclei
c. protons

b. electrons
d. electromagnetic radiation

10. Which hydrogen isotope contains one neutron?

a. protium
c. tritium

b. deuterium
d. none

11. Which is an isotope of carbon–12?

a. $^{13}_{6}C$

b. $^{12}_{5}B$

c. $^{13}_{7}N$

d. $^{12}_{7}N$

12. James Chadwick discovered the neutron in 1932 by bombarding beryllium with alpha particles:

$$^{9}_{4}Be \ + \ ^{4}_{2}He \ \longrightarrow \ ^{1}_{0}n \ + \ ?$$

A neutron was produced; what is the other product?

a. $^{11}_{4}Be$

b. $^{3}_{2}He$

c. $^{13}_{6}C$

d. $^{12}_{6}C$

13. Carbon –14 undergoes beta decay:

$$^{14}_{6}C \longrightarrow ^{0}_{-1}e + ?$$

What is the other product?

a. $^{13}_{6}C$

b. $^{14}_{5}B$

c. $^{13}_{7}N$

d. $^{14}_{7}N$

14. Which of the following are characteristic of a nuclear reaction (more than one answer is possible)?

a. Mass is conserved
b. energy is conserved
c. mass and energy is conserved
d. neither mass nor energy is conserved

15. Iréne Curie and her husband, Frédéric Joliot, discovered artificial radioactivity in 1933 by bombarding aluminum with alpha particles. The new radioactive element was

$$^{27}_{13}Al + ^{4}_{2}He \longrightarrow ^{30}_{15}P + ?$$

What is the other product of the reaction?

a. $^{1}_{2}He$

b. $^{1}_{0}n$

c. $^{0}_{+1}e$

d. $^{0}_{-1}e$

16. Which isotope will undergo nuclear fission when bombarded with neutrons?

a. $^{3}_{1}H$

b. $^{238}_{92}U$

c. $^{235}_{92}U$

d. $^{4}_{2}H$

17. Binding energy is caused by

a. mass being produced during a nuclear change
b. mass being lost during a nuclear change
c. charge being produced by a nuclear change
d. charge being lost in a nuclear change

18. Which isotope(s) is/are fuel for nuclear fusion in the hydrogen bomb?

   a. $^{235}_{92}U$

   b. $^{238}_{92}U$

   c. $^{1}_{1}H$ and $^{2}_{1}H$

   d. $^{2}_{1}H$ and $^{3}_{1}H$

19. What kind of nuclear reactor is the sun?

   a. neutron
   c. fission

   b. isotope
   d. fusion

20. You are exposed to radiation

   a. because of atmospheric testing of nuclear weapons
   b. when your teeth are X-rayed by a dentist
   c. simply because you live on the Earth
   d. for all of these reasons

21. How much energy in the United States is produced by nuclear energy?

   a. 5%
   c. 20%

   b. 10%
   d. 50%

22. For which of the following would N be an improper symbol?

   a. $^{14}_{8}X$

   b. $^{14}_{7}X$

   c. $^{15}_{7}X$

   d. $^{13}_{7}X$

23. Which type of radiation is a stream of charged particles?

   a. alpha rays
   c. gamma rays

   b. x-rays
   d. cosmic rays

24. Which isotope is particularly useful for both diagnostic and therapeutic work with the thyroid gland?
   a. cobalt-60
   c. technetium-99m

   b. iodine-131
   d. tritium

25. The isotope that has nearly ideal properties for a large number of diagnostic scanning uses, including brain scans, is
   a. I-131
   c. U-235

   b. Tc-99^{m}
   d. Co-60

26. Which process does this equation illustrate?

$$^{2}_{1}H + ^{2}_{1}H \longrightarrow ^{4}_{2}He$$

    a. fission   b. fusion

    c. radioactivity   d. none of these

27. A fission nuclear reaction requires

    a. high temperatures   b. low temperatures

    c. critical mass   d. supercritical mass

28. A fusion nuclear reaction requires

    a. high temperatures   b. low temperatures

    c. critical mass   d. supercritical mass

29. Which particle causes the most damage internally to the body?

    a. alpha   b. beta

    c. delta   d. gamma

30. A chain reaction requires

    a. high temperatures   b. low temperatures

    c. critical mass   d. all of the above

True or False

___ 31. Chemical bond energies are only slightly smaller than nuclear bond energies.

___ 32. Radioactive fallout can result in the incorporation of some radioactive isotopes into the food chain.

___ 33. Fission reactions release more energy than fusion reactions.

___ 34. The least penetrating of the three kinds of radioactive decay are gamma rays.

___ 35. Alpha particles can be stopped by a sheet of paper.

___ 36. One way to decrease exposure to radiation is to move farther from the source.

___ 37. Nuclear reactions can be slowed down by lowering the temperature

ANSWERS

Additional Problems

1. a. $^{234}_{91}\text{Pa}$

   b. $^{218}_{84}\text{Po}$

   c. $^{4}_{2}\text{He}$

2. After 1 minute, 200 μg; after 2 minutes, 100 μg; after 4 minutes, 25 μg.
3. 72,900 years (three half-lives)

SELF-TEST

1. d	7. a	13. d	19. d	25. b	31. F	37. F
2. b	8. b	14. c	20. d	26. b	32. T	
3. c	9. a	15. b	21. c	27. c	33. F	
4. b	10. b	16. c	22. a	28. a	34. F	
5. b	11. a	17. b	23. a	29. a	35. T	
6. a	12. d	18. d	24. b	30. c	36. T	

Chemical Bonds

The Ties that Bind

KEY TERMS

binary compounds
bonding pair
crystal
condensation
covalent bond
dispersion forces
dipole interactions
double bond
electron dot symbols
electron dot formulas
electronegativity
free radicals

freezing
ions
ionic bonds
homogeneous
hydrogen bond
nonbonding pairs
melting point
nonpolar covalent bonds
octet rule
polar covalent bonds
polar molecule

polyatomic ions
single bond
solvent
solute
solution
triple bond
vaporization
valence electrons
VSEPR theory

CHAPTER SUMMARY

Introduction
    A.   Chemical bonds hold atoms and ions together.
       1.   Nature of bonding within the molecule determines the forces between molecules.
       2.   Bonds determine the shape of the molecule.

5.1   The Art of Deduction: Stable Electron Configurations
    A.   Atoms can gain or lose electrons to form electron configurations resembling those of the noble gases.
    B.   When atoms gain or lose electrons and acquire a charge, they are called ions.

5.2   Electron Dot Structures
    A.   Electron dot (electron dot symbols are also called Lewis symbols) are structural symbols, and they are a useful way to represent atoms or ions.
       1.   The chemical symbol represents the core (nucleus plus inner electrons).
       2.   Valence electrons are represented by dots.
       3.   Group number of A group elements gives the number of valence electrons.
          a.   Na (IA)—one valence electron Cl (VIIA)—seven valence electrons
    B.   Chemical Symbolism
       1.   Electron dot symbols are more convenient to use than energy level diagrams.
          a.   For the "A" group elements, the number of valence electrons is equal to the group number.

5.3   Sodium Reacts with Chlorine: The Facts
A.   Sodium is a very soft, reactive metal. Chlorine is a poisonous, greenish yellow, reactive gas.
B.   When sodium metal is added to chlorine gas, a violent reaction takes place and produces a white, stable, water-soluble solid known as sodium chloride (table salt).

5.4   Sodium Reacts with Chlorine: The Theory
A.   Sodium reacts with chlorine by giving it an electron.
B.   The ion products of this reaction are stable because both ions have stable noble gas-type electron configurations. The ions have opposite charges, so the ions are attracted to each other, forming ionic bonds.
   1.   Compounds with ionic bonds often form crystalline solids that have well-defined regular shapes.

5.5   Using Electron Dot Symbols: More Ionic Compounds
A.   Potassium (same family as sodium) reacts with chlorine the same way sodium does.
B.   Potassium reacts with bromine (same family as chlorine) the same way it reacts with chlorine.
C.   Magnesium (Group IIA) reacts with oxygen (Group VIA) by giving up two electrons.
D.   Metallic elements of Groups IA, IIA, and IIIA react with nonmetallic elements in Groups VA, VIA, and VIIA to form stable crystalline solids.
   1.   Metals give up electrons forming (+) ions.
   2.   Nonmetals gain electrons forming (–) ions.
   3.   The attraction of (+) ions for (–) ions is the basis for ionic bonds. The resulting compound is called a salt.
E.   Octet Rule
   1.   All noble gases have an octet (8) of valence electrons except He, which has two.
   2.   Atoms lose or gain electrons to follow the octet rule according to their group number:
         Group IA lose one electron to form +1 cations
         Group IIA lose two electrons to form +2 cations
         Group VIA gain two electrons to form –2 anions
         Group VIIA gain one electron to form –1 anions

5.6   Formulas and Names of Binary Ionic Compounds
A.   Names and Symbols for Simple Ions
   1.   To name simple cations, add "ion" to the name of the parent element.
      a. Na^+, sodium ion.
   2.   To name simple anions, change the ending of the name of the element to "-ide" and add "ion."
      a. Cl^-, chloride ion.
   3.   Elements in the "B" groups on the Periodic Table can form more than one ion. In those cases,
B.   Formulas and Names for Binary Ionic Compounds
   1.   The "crossover" method is used to determine the numbers of cations and anions needed for the correct formula of an ionic compound. For aluminum sulfide $Al^{3+} + S^{2-} \rightarrow Al_2S_3$

5.7   Covalent Bonds: Shared Electron Pairs
A.   Elements that are unable to transfer electrons completely between them achieve electronic configurations by sharing pairs of electrons.
B.   A bond formed by sharing a pair of electrons is called a covalent bond. Covalently bonded atoms, except hydrogen, seek an arrangement that surrounds them with eight electrons (octet rule).
   1.   The two shared electrons forming the bond are called a bonding pair.
   2.   The electrons that stay with each atom and are not shared are the nonbonding pairs.
   3.   Covalent bonding can be symbolized in several ways.
C.   Multiple Bonds
   1.   Atoms can share more than one pair of electrons resulting in:

  a. Double bonds (two pairs of electrons shared between two atoms).

  b. Triple bonds (three pairs of electrons shared between two atoms).

 D. Names of Covalent Compounds

  1. When two nonmetals form a covalently bonded compound, prefixes are used to indicate the number of atoms of each element in the molecule.

Prefix	Number
mono-	1
di-	2
tri-	3
tetra-	4

  2. The prefix mono- is omitted; SO_2 is sulfur dioxide (not monosulfur dioxide).

5.8 Unequal Sharing: Polar Covalent Bonds

 A. Polar covalent bond: electron pair is unequally shared between the two bonded atoms.

  1. The element in a polar covalent bond with a greater attraction for electrons is more electronegative than the other element.

  2. Polar covalent bonded molecules are represented by (δ^+) over the least electronegative partner and (δ^-) over the more electronegative partner.

   a. δ^+ δ^-

    H-Cl

 B. Electronegativity is a measure of the attraction of an atom in a molecule for a pair of shared electrons

  1. Nonmetals are more electronegative than metals. Fluorine (upper right corner of the periodic table) is the most electronegative element, and metals like cesium (lower left) are the least electronegative.

5.9 Polyatomic Molecules: Water, Methane, and Ammonia

 A. To calculate the number of covalent bonds a nonmetallic element (Groups VIA-VIIA) will form, subtract the group number from 8.

  1. Oxygen in Group VIA: $8 - 6 = 2$ bonds

 B. Water

  1. The molecular formula for water is H_2O.

  2. The electron dot formula for water is

  3. Water is shown as a bent molecule, rather than a linear molecule, to best demonstrate its polar nature.

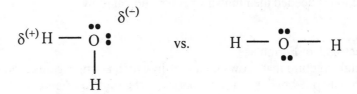

C. Ammonia
   1. The molecular formula for ammonia is NH_3.
   2. The electron dot formula for ammonia is

   3. The shape is pyramidal to best explain its polar nature.

D. Methane
   1. The molecular formula for methane is CH_4.
   2. It forms a tetrahedron with the H bond pointing to each corner of the geometric figure.

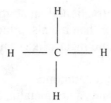

5.10 Polyatomic Ions
   A. Polyatomic ions are groups of atoms that remain together during most chemical reactions. They are bonded to each other by covalent bonds and have an overall charge.
   B. A number of polyatomic ions are so common that they have been given names.

   1. | Examples | Name | Formula | Name | Formula |
       |---|---|---|---|---|
       | | Ammonium ion | NH^{4+} | Nitrate ion | NO_3^- |
       | | Carbonate ion | CO_3^{2-} | Phosphate ion | PO_4^{3-} |
       | | Hydroxide ion | OH^- | Sulfate ion | SO_4^{2-} |

   C. In balanced formulas, a subscript outside the parentheses surrounding a polyatomic ion indicates that the entire polyatomic ion is needed the number of times indicated by the subscript.

5.11 Rules for Writing Electron Dot Formulas
   A. Construct a skeletal structure that shows the order in which atoms are attached to each other.
      1. H atoms form only one bond, so they cannot be in the middle of a skeletal structure.
         a. H atoms are often bonded to C, N, or O.
      2. In polyatomic molecules, the central atom is surrounded by more electronegative elements.
   B. Calculate the total number of valence electrons by summing the number of valence electrons from each atom.
      1. In polyatomic anions, the number of the negative charge is added to the valence electron total.
      2. In polyatomic cations, the number of the positive charge is subtracted from the valence electron total.
   C. Draw lines to represent bonds between atoms in the skeletal structure.
      1. Subtract two electrons from the total number of valence electrons for each bond drawn.

D. Assign the remaining valence electrons so that each atom is surrounded by eight electrons (except H and He, which are surrounded by two electrons).
    1. If there are not enough valence electrons to satisfy each atom, assign a double or triple bond between atoms and redistribute remaining valence electrons.

5.12 Exceptions to Octet Rule
  A. Odd-electron molecules: Free radicals
    1. Molecules with odd numbers of valence electrons cannot satisfy the octet rule. Such molecules are free radicals.
      a. Many free radicals are highly reactive.
      b. Nitrogen oxides are examples of stable free radicals found in smog.
  B. Molecules with incomplete octets
    1. Some molecules exist with fewer than eight electrons on the central atom.
      a. Boron and beryllium are examples of central atoms that can exist with fewer than eight electrons.
  C. Molecules with expanded octets
    1. Elements in the third period use the third main energy level, which can hold up to 18 electrons instead of eight. Their atoms, therefore, can have more than eight valence electrons when they serve as central atoms in molecules.

5.13 Molecular Shape: VSEPR Theory
  A. VSEPR is used to predict the three-dimensional arrangement of atoms around the central atom.
    1. Electron pairs are arranged around a central atom to minimize repulsion.
      a. Example: In a molecule with two bonds and no nonbonding pairs around the central atom, the farthest that the two bonds can get is 180° forming a linear molecule.
  B. To determine molecular shape:
    1. Draw an electron dot structure.
    2. Count the number of bonds and nonbonding pairs attached to central atom.
      a. A multiple bond counts as one bond.
    3. Draw the shape as if all bonds are single bonds.
    4. Erase nonbonding pairs.

5.14 Shapes and Properties: Polar and Nonpolar Molecules
  A. Diatomic molecules
    1. A molecule is polar if its bonds are polar.
    2. A molecule is nonpolar if its bonds are nonpolar.
  B. Other polar molecules (consider the orientation of the bonds to determine if molecule is polar.
    1. Water: A bent molecule
      a. Water acts like a dipole (a molecule with a positive and negative end).
        The water molecule must be bent; otherwise the charge on the bonds would cancel each other.
      b. According to VSEPR, the two bonds and two nonbonding pairs of electrons on water's oxygen should form a tetrahedron shape. However, the nonbonding pairs occupy a larger volume and push the two bonds closer together forming an angle of 104.5° instead of the tetrahedral angle of 109.5°.
    2. Ammonia: A pyramidal molecule
      a. VSEPR predicts a tetrahedral shape but the nonbonding pair on nitrogen pushed the bonds closer together forming an angle of 107° rather than 109.5°.
      b. Shape is pyramidal.
  C. Nonpolar molecules
    1. Methane: A tetrahedral molecule
      a. VSEPR predicts a tetrahedral shape. The four bonds (nonbonding pairs) form angles of 109.5°.
      b. Slight bond polarities are canceled by the symmetrical shape of the molecule.
      c. The molecule is nonpolar.

5.15 Intermolecule Forces
   A. Compounds that have intermolecular forces are either solids or liquids. Compounds without significant intermolecular forces are gases.
      1. Solids can be changed to liquids by adding energy; the temperature at which this happens is called the melting point.
      2. Liquids can be changed to gases by adding energy (vaporization) or frozen to solids by removing energy (freezing).
      3. Gases can be condensed to the liquids by removing energy (condensation).
   B. Dipole forces
      1. Unsymmetrical molecules containing polar bonds are dipoles with centers of partially negative and positive charges.
      2. Polar molecules attract one another as the positive end of one molecule interacts with the negative end of another molecule. Dipole forces are weaker than ionic bonds but stronger than forces between nonpolar molecules of comparable size.
   C. Hydrogen bonds
      1. Compounds containing H attached to small, electronegative elements such as N, O, or F exhibit stronger intermolecular attractive forces than would be expected on the basis of dipolar forces alone. These forces are called hydrogen bonds.
      2. A hydrogen bond is much weaker than a covalent bond. The hydrogen bond is an interaction of the partially positive hydrogen of the donor molecule with the lone pair of nonbonding electrons of the F, O, N of the acceptor molecule.
         a. Hydrogen bonds are usually represented by dotted lines.
   D. Dispersion forces
      1. Nonpolar compounds experience attractive intermolecular forces due to momentary induced dipoles arising from the motions of electrons around the nuclei of atoms in the compound.
      2. These transient attractive forces, called dispersion forces, are fairly weak but are present in all molecules and increase as the size and number of electrons in the molecule increase.
      3. Dispersion forces can be substantial between large molecules, such as those in polymers.
   E. Forces in solutions
      1. A solution is a homogeneous mixture of two or more substances.
      2. The solute is the substance being dissolved (the minor component).
      3. The solvent is the substance doing the dissolving (the major component).
      4. "Like dissolves like."
         a. Nonpolar substances dissolve best in nonpolar solvents, polar substances in polar solvents.
      5. Salts dissolve in water because the ion-dipole forces overcome the ion-ion attractions.

DISCUSSION

Electrons are the glue that holds atoms together. Atoms achieve the stable octet of electron configuration by transferring electrons (ionic bonds) from one to another or by sharing electrons (covalent bonds). Since only valence electrons (electrons outside of the filled electron shells) are used, an abbreviated notation called the electron dot symbol is used. The group number gives the number of valence electrons. For example, NaCl is formed from unstable Na metal and reactive Cl_2 gas. Electron configurations or electron dot symbols explain the properties of Na, Cl_2, and NaCl. Covalent compounds are formed when the atoms share electrons. There are some important exceptions to the octet rule: reactive free radicals, small atoms and large atoms with more than 8 electrons that use d orbitals. The VSEPR theory is used to predict shapes of molecules by looking at the number of electron-pairs around the atoms. The shape (and polarity) of molecules can be used to explain the states of matter. Dipole forces, hydrogen bonds, dispersion forces, and forces in solutions are the types of bonding between molecules.

Many of the problems at the end of the chapter are drills. They have you practice drawing ions, putting together molecules or naming compounds. This practice should help firmly establish the rules governing chemical structure in your mind. In case it doesn't, here is some additional help.

First, how do you know whether a compound is ionic or covalent? Just follow these rules:
1. When hydrogen bonds with other nonmetals, the compounds are covalent.
    Examples: H_2O, NH_3 and CH_4.
2. When hydrogen bonds with Group IA or IIA metals, the compounds are ionic.
    Examples: NaH, CaH_2, and LiH.
3. When Group IA or IIA metals bond with Group VA, VIA, or VIIA nonmetals, the compounds are ionic.
    Examples: NaCl, CaF_2, K_2S, and Na_3N.
4. When nonmetals bond with other nonmetals, the compounds are covalent.
    Examples: CO_2, CCl_4, and PCl_3.

Second, if the compound is covalent, how do you know whether it is polar or nonpolar?
1. If both (or all) of the atoms in the molecule are the same, the compound is nonpolar.
    Examples: O_2, N_2, Br_2, and S_8
2. If the atoms in the molecule are not the same, the bonds are polar.
    Examples: NO, PCl_3, SO_2, H_2, and HCl.
3. To decide if the molecule is polar or nonpolar, you must determine its shape.
    a. If the shape is symmetrical, the <u>molecule</u> is <u>nonpolar</u>.
    b. If the shape is not symmetrical, the <u>molecule</u> is <u>polar</u>.

CHAPTER OBJECTIVES

(You should be able to...)

1. Recognize compounds that have ionic bonds. (The elements are far apart in the periodic table; examples are NaCl, K_2O, and CaF_2.)

2. Recognize compounds that have polar covalent bonds. (The elements are closer together in the periodic table but have different electronegativities; examples are NO, PCl_3, and CS_2.)

3. Recognize molecules that have nonpolar covalent bonds. (These molecules are composed of the same kind of atom; examples are N_2, Cl_2, and S_8.)

4. Determine simple combining ratios of atoms.

5. Given the molecular formula, construct structural formulas for simple covalent compound. (Electrons are transferred or shared until each has eight electrons in the outer shell; hydrogen and helium are exceptions, requiring only two.)

6. Determine the number of bonds all elements in Groups IVA to VIIA will form. (To determine charges: for a metal, the charge is positive and equal to the group number; for nonmetals, the charge is negative and equal to eight minus the group number.)

7. Use the VSEPR theory to predict the shape of simple molecules.

8. Write electron dot symbols for any of the first 20 elements. (The number of dots representing valence electrons is equal to the group number.)

9. Write symbols with charges, for simple ions.

10. Recognize that ionic bonds are formed by the transfer of electrons.

11. Be able to name binary ionic compounds or write their formula from the name.

12. Be able to name common polyatomic ions given their formula or write the formula given the name.

13. Recognize that covalent bonds are formed by the sharing of electrons.

14. Recognize that polar covalent bonds are formed by the unequal sharing of electrons.

15. Be able to name binary covalent compounds given their formula or write their formula given their name.

16. Recognize that nonpolar covalent bonds are formed by the equal sharing of electrons.

17. Determine the type of bonding between molecules: dipole, H-bonds, dispersion force as solutions.

18. Write electron dot formulas for simple ionic and covalent compounds.

19. Define or identify each of the key terms in this chapter.

EXAMPLE PROBLEMS

1. Give the electron dot formula for H_2S

The parts are
$$H^{\bullet}, H^{\bullet}, \text{ and } {}^{\bullet}\overset{\bullet\bullet}{\underset{\bullet}{S}}{}^{\bullet\bullet}$$

One of the unpaired electrons of S pairs with the electron of an H to form a bond. The second single electron of S combines with the single electron of the remaining hydrogen.

$$H \overset{\bullet\bullet}{\underset{\bullet\bullet}{:S:}} H$$

Voila!

2. Give the electron dot formula for CH_4S

$$^{\bullet}\overset{\bullet}{\underset{\bullet}{C}}{}^{\bullet} \quad H^{\bullet} \quad H^{\bullet} \quad H^{\bullet} \quad H^{\bullet} \quad {}^{\bullet}\overset{\bullet\bullet}{\underset{\bullet}{S}}{}^{\bullet\bullet}$$

Step 1

Step 2 Hydrogen is always in an outer position.

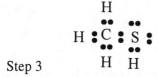

Four bonds still to be formed

H• H• H• H•

Four bonds to be formed by these atoms with this combination of atoms

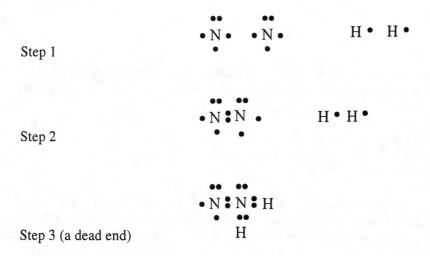

Step 3

3. Write the electron dot formula for N_2H_2

Step 1

•N• •N•   H• H•

Step 2

•N:N•   H•H•

Step 3 (a dead end)

•N:N:H
H

The nitrogen on the left still must form two bonds, and there are no other atoms left to bond with.

Step 4 (a retreat to Step 2)

•N:N•   H•H•

Step 5 (a different placement of the H's)

H:N:N:H

Each nitrogen still must form one more bond, so they bond to each other again.

Step 6   H:N:N:H

4.   Is MgS ionic or covalent?

Mg is in Group IIA, S is in Group VIA. These elements are far apart on the periodic table; the compound is ionic.

5.   What is the correct formula for calcium chloride?

Calcium (Ca) is in Group IIA and has two electrons to give. Chlorine (Cl) is in Group VIIA and can accept only one electron. Therefore, there must be two chlorine atoms for each calcium atom: the formula is $CaCl_2$.

ADDITIONAL PROBLEMS

1. Compounds are formed from the following sets of elements. Indicate whether the compound would be ionic or covalent. (Note that you're not being asked to draw the compounds, just to evaluate their tendency to form ionic or covalent bonds.)

    a. Mg and O   b. F and Ca   c. Li and S   d. Br and Cl
    e. Na and H   f. S and H   g. Ba and Br   h. C and O
    i. N and O   j. Rb and F

2. You should be able to draw electron dot symbols for any element in the A groups (IA, IIA, etc.). To do this, write the symbol for the element and surround it with dots representing the valence (outermost) electrons. The number of valence electrons is given by the group number. For practice, draw electron dot structures for the atoms of these elements.

    a. barium   b. carbon   c. xenon   d. nitrogen
    e. silicon   f. hydrogen   g. potassium   h. chlorine
    i. sulfur   j. boron

3. Electron dot structures for ions are drawn either by adding electron dots to complete the octet or by removing electron dots to empty the outermost level. Electrons are added to elements with five or more valence electrons; they are subtracted from elements with three or fewer valence electrons. For the following elements, how many electron dots would you add to or remove from the electron dot symbols of the atoms to form the ions?

    a. barium   b. chlorine   c. nitrogen   d. iodine
    e. potassium   f. magnesium   g. sulfur   h. aluminum

4. The charge is written to the upper right of the electron dot symbol for an ion. The charge is equal to the number of electrons added to or removed from the neutral atom to form the ion. The charge is positive if electrons are removed and negative if electrons are added. Write electron dot symbols for the ions that are formed from the eight elements listed in problem 3.

SELF-TEST

1. The electron dot symbol for a potassium atom is

   a. K•

   b. K :

   c. :K:

   d. :K: (with dots above and below)

2. The correct electron dot symbol for the beryllium atom is

   a. Be•

   b. Be• (with dot below)

   c. :Be:

   d. :Be: (with dot below)

3. The correct electron dot symbol for the carbon atom is

   a. C:

   b. •C• (with dots above and below)

   c. :C: (with dots above)

   d. :C: (with dots above and below)

4. The correct electron dot structure for a Group IV a atom is

   a. X (with dot above)

   b. X• (with dot above)

   c. X• (with dots above and below)

   d. •X• (with dots above and below)

5. Sodium ions (Na⁺) differ from sodium atoms (Na) in that the ions have one
   a. more proton than the atom
   b. less proton than the atom
   c. more electron than the atom
   d. less electron than the atom

6. Chloride ions (Cl⁻) differ from chlorine atoms (Cl) in that the ions have one

    a. more proton than the atom
    b. less proton than the atom
    c. more electron than the atom
    d. less electron than the atom

7. Group III atoms lose

    a. 1 electron   b. 2 electrons
    c. 3 electrons   d. 4 electrons

8. Which substance would most likely have ionic bonds?

    a. HCl b. Na_2O
    c. Cl_2 d. N_2O

9. Group II atoms

    a. gain 1 electron   b. gain 2 electrons
    c. lose electrons   d. lose 2 electrons

10. Which substance has polar covalent bonds?

    a. CaF_2 b. F_2
    c. HF d. H_2

11. The correct electron dot structure of a Group VIIA atom, X, is

    a.   • X •   b.   • X :

    c.   : X :   d.   • X •

12. Metals tend to _____ electrons while non-metals tend to _____ electrons.

    a. gain, gain   b. gain, lose
    c. lose, lose   d. lose, gain

13. Which substance has nonpolar covalent bonds?

    a. F_2 b. HF
    c. OF_2 d. H_2O

14. Sodium (Na) reacts with sulfur (S) to form

    a. NaS b. Na_2S
    c. Na_3S d. NaS_2

15. Calcium reacts with chlorine to form

    a. $CaCl$ b. $CaCl_2$

    c. Ca_2Cl d. Ca_2Cl_2

16. The simplest compound of nitrogen (N) with hydrogen (H) is

    a. NH b. NH_2

    c. NH_3 d. NH_4

17. The simplest compound of carbon with hydrogen is

    a. CH b. CH_2

    c. CH_3 d. CH_4

18. The most likely charge on an aluminum (Al) ion is

    a. 1+   b. 2+

    c. 3+   d. 3–

19. The most likely charge on an F ion is

    a. 1+   b. 1–

    c. 2+   d. 3–

20. Mg^{2+} is the symbol for

    a. tin (II) ion   b. magnesium ion

    c. zinc ion   d. zirconium ion

21. Fe^{3+} is the symbol for

    a. fermium ion   b. fluoride ion

    c. iron (III) ion   d. indium ion

22. The symbol for nitride ion is

    a. NH_4^+ b. NO^{2-}

    c. NO_3^- d. N^{3-}

23. Cu^{2+} is called

    a. cobalt (II) ion   b. chromium (II) ion

    c. copper (II) ion   d. chloride ion

24. Lithium nitrate is

    a. $LiNO_2$ b. $LiNO_3$

    c. $LiNO_4$ d. Li_3N

25 Aluminum chloride is

    a. $AlCl$ b. $AlCl_2$

    c. $AlCl_3$ d. $AlClO_2$

26. Calcium phosphate is

    a. $CaPO_4$ b. Ca_2PO_4

    c. $Ca_2(PO_4)_3$ d. $Ca_3(PO_4)_2$

27. The name of the CN⁻ ion is

    a. carbonate   b. carbonite

    c. cyanate   **d.** cyanide

28. Copper (I) oxide is

    a. CuO **b.** Cu_2O

    c. CuO_2 d. CuO_3

29. Iron (II) sulfate is

    a. $FeSO_4$ b. Fe_2SO_4

    c. $Fe(SO_4)_2$ d. $Fe_2(SO_4)_3$

30. Sodium nitrite is

    a. Na_3N b. $NaNH_2$

    c. $NaNO_2$ d. $NaNO_3$

31. Copper (I) ion is

    a. Cu^+ b. Cu^{2+}

    c. Cu_2^+ d. CuO^+

32. The formula for lithium oxide is

    a. LiO b. LiO_2

    c. Li_2O **d.** Li_2O_3

33. The formula for copper (II) fluoride is

    a. CuF **b.** CuF_2

    c. Cu_2F d. CU_2F_2

34. The formula Cu_3N represents

    a. copper (I) nitride   b. copper (II) nitride

    c. copper (III) nitride   d. copper nitrite

35. The formula FeO represents

    a. iron (I) oxide   b. iron (II) oxide

    c. iron (III) oxide   d. iron (IV) oxide

36. PCl_3 is

    a. phosphorous chloride   **b.** phosphorus trichloride

    c. triphosphorus trichloride   d. triphosphorus chloride

37. The formula for sulfurhexafluoride is

    a. S_6F **b.** SF_6

    c. S_6F_6 d. SF_2

38. The formula for carbon tetrabromide is

    a. C_4B b. C_4Br

    c. CB_4 **d.** CBr_4

39. N_2O_4 is

    a.   nitrogen oxide   b.   dinitrogen oxide

    c.   nitrogen tetraoxide   (d.) dinitrogen tetraoxide

40. The shape of the ammonia molecule (NH_3) is best described as

    a.   linear   b.   bent

    (c.) pyramidal   d.   triangular

41. Which substance has ionic bonds?

    a. O_3 b. SO_2

    c. H_2S (d.) Na_2S

42. Which substance has covalent bonds?

    a.   NaCl   b.   CaO

    (c.) NO   d. K_2S

43. Which is the correct electron dot formula for HCN?

    a. $\bullet\!\bullet$ H $\vdots$ C $\vdots$ N   b. $\bullet\!\bullet$ H $\vdots$ C $\vdots\vdots$ N $\vdots$

    c. H $\vdots$ C $\vdots\vdots$ N $\vdots$ (d.) H $\vdots$ C $\vdots\vdots\vdots$ N $\vdots$

44. The methane molecule (CH_4) is

    a.   linear   b.   bent

    c.   pyramid   (d.) tetrahedral

45. The water molecule is

    (a.) bent   b.   linear

    c.   pyramidal   d.   tetrahedral

46. Select the correct structure for acetylene (C_2H_2).

    a. H — C — C — H   (b.) H — C ≡ C — H

    c. H ≡ C ≡ C ≡ H   d. H — C ═ C — H

47. What is the shape of the H_2S molecule?

    (a.) bent   b.   linear

    c.   pyramidal   d.   tetrahedral

48. In nearly all molecules, how many bonds does carbon form?

    a.   2   b.   3

    (c.) 4   d.   5

49. Which is the correct structure for CH₃F?

    a. C - H - H - H - F

    b. H - C - H - F
              |
              H

    c. H - C - F - H
              |
              H

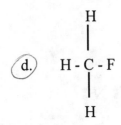

    d. H - C - F
              |
              H

50. Which substance is made up of polar covalent molecules?

    a. S_8 b. Na_2S

    c. H_2S d.   NaF

51. The correct electron dot formula for C_2H_2 is

    a.   H ⦂ C ⦂ C ⦂ H   b. ⦂C ⦂ C ⦂ H ⦂ H

    c.   H ⦂ C ⦂⦂ C ⦂ H   d. H ⦂ C ⦂⦂ C ⦂ H

52. Select the correct structure for methylamine (CH_5N).

a.

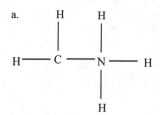

b.

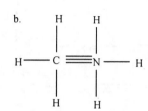

c.

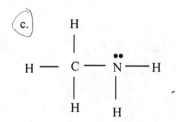

d.

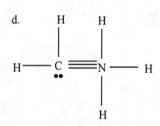

53. A chemical bond formed by the equal sharing of electrons is called

    a. ionic
    c. polar covalent
    b. nonpolar covalent
    d. coordinate

54. Which is the best formula for magnesium oxide?

    a. MgO
    c. Mg_2O
    b. MgO_2
    d. Mg_2O_2

55. Which is the correct formula for a compound of boron and sulfur?

    a. BS
    c. B_2S_3
    b. B_2S
    d. B_3S_2

56. Which is the correct formula for a compound of aluminum and nitrogen?

    a. AlN
    c. AlN_3
    b. Al_3N
    d. Al_3N_3

57. The correct electron dot formula for carbon dioxide is

a. :C : O : C :

b. :O : O : C :

c. :O :: C :: O :

d. :O :: C :: O :

58. Which is the most reasonable structural formula for C_2F_2?

a. F —— C —— C —— F

b. C —— F —— F —— C

c. F ≡≡ C —— C ≡≡ F

d. F —— C ≡≡ C —— F

59. The correct electron dot formula for HNO is

a. H : N :: O :

b. H : N :: O :

c. H : O :: N :

d. :H : O   N :

60. A reasonable structure for C_2H_3OCl is

a.

b.

c.

d.

64

61. Which state of matter is characterized by having molecules close together but moving at random?

   a. gas
   c. solid
   b. liquid
   d. all of these

62. Which represent the best arrangement for a pair of dipoles?

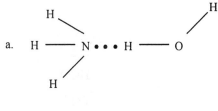

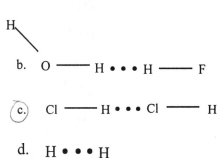

63. Which of the following correctly illustrates hydrogen bonding?

   a.
   H — N • • • H — O
   (with H atoms branching from N and O)

   b.
   O — H • • • H — F
   (with H branching from O)

   c. Cl — H • • • Cl — H

   d. H • • • H

64. Which is the most highly ordered arrangement of matter?

   a. gas
   c. solid
   b. liquid
   d. all are the same

65. In general, which is the strongest type of bonding?

   a. ionic bonding
   c. hydrogen bonding
   b. dipole interactions
   d. dispersion forces

66. Which solute is most likely to dissolve in carbon tetrachloride (a nonpolar covalent compound)?

   a. Br_2
   c. NaCl
   b. H_2O
   d. NaBr

67. As the temperature of a solid increases

   a. the particles move closer together
   c. the particles move faster
   b. the particles lose electrons
   d. the particles change color

68. Which physical state has the weakest attractive forces between particles?

    a. gases   b. liquids

    c. solids   d. they all have the same forces

69. The weakest forces between particles are

    a. H-bonding   b. dipole forces

    c. dispersion forces   d. they are all equal

70. Which dissolves best in water?

    a. CH_4 b. H_2

    c. O_2 d. KF

Matching

Match the name with the formula of the polyatomic ion

f 1. Ammonium ion   a. HCO_3^-

g 2. Acetate ion   b. PO_4^{3-}

e 3. Sulfate ion   c. H_3O^+

b 4. Phosphate ion   d. NO_3^-

c 5. Hydronium ion   e. SO_4^{2-}

h 6. Hydroxide ion   f. NH_4^+

d 7. Nitrate ion   g. $HC_2H_3O_2^-$

a 8. Bicarbonate ion   h. OH^-

ANSWERS

Additional Problems

1. Ionic: a, b, c, e, g, j Covalent: d, f, h, i

2. a. Ba·   b. ·C·   c. :Xe: d. ·N: e. ·Si·

   f. H·   g. K·   h. :Cl·   i. ·S: j. ·B·

3. a. remove two   b. add one   c. add three   d. add one
   e. remove one   f. remove two   g. add two   h. remove three

4. a. Ba^{2+} b. $\left[:\overset{..}{\underset{..}{Cl}}: \right]^{-}$ c. $\left[:\overset{..}{\underset{..}{N}}: \right]^{3-}$ d. $\left[:\overset{..}{\underset{..}{I}}: \right]^{-}$

   e. K^{+} f. Mg^{2+} g. $\left[:\overset{..}{\underset{..}{S}}: \right]^{2-}$ h. Al^{3+}

Self-Test

1. a	12. d	23. c	34. a	45. a	56. a	67. c
2. b	13. a	24. b	35. b	46. b	57. c	68. a
3. b	14. b	25. c	36. b	47. a	58. d	69. c
4. d	15. b	26. d	37. b	48. c	59. a	70. d
5. d	16. c	27. d	38. d	49. d	60. b	
6. c	17. d	28. b	39. d	50. c	61. b	
7. c	18. c	29. a	40. c	51. c	62. a	
8. b	19. b	30. c	41. d	52. c	63. c	
9. d	20. b	31. a	42. c	53. b	64. c	
10. c	21. c	32. c	43. d	54. a	65. a	
11. c	22. d	33. b	44. d	55. c	66. a	

Matching

1. f	2. g	3. e	4. b	5. c
6. h	7. d	8. a		

CHAPTER

6

Chemical Accounting

Mass and Volume Relationships

KEY TERMS

Aqueous solutions
Avogadro's hypothesis
Avogadro's number
binary compounds
Boyle's Law
Charles's Law
chemical equation
concentrated solution

dilute solution
formula mass
Ideal gas law
Kinetic Molecular Theory
Law of combining volumes
molarity
molar mass
molar volume

mole (mol)
molecular mass
percent by mass
percent by volume
products
reactants
standard temperature and pressure (STP)
stoichiometry

CHAPTER SUMMARY

6.1 Chemical Sentences: Equations
   A. A chemical equation is a shorthand notation for describing chemical change as the reactants (shown on the left of the arrow) are converted to products (shown on the right of the arrow)
      1. R $\rightarrow$ P.
   B. Chemical equations involve electron rearrangements as chemical bonds are broken and formed. The nuclei of all atoms remain unchanged during chemical reactions (as opposed to nuclear reactions).
   C. Balancing chemical equations—because atoms are converted in chemical reactions, the same number of each kind of atom must appear on both sides of a balanced chemical equation.

$$2\,H_2 + O_2 \rightarrow 2\,H_2O$$

   1. When counting atoms, multiply the coefficients in front of formulas times the subscripts for a given atom in the formula.
   2. Polyatomic ions are treated as units whenever possible in balancing chemical equations.

6.2 Volume Relationships in Chemical Equations
   A. Law of combining volumes: When all measurements are made at the same temperature and pressure, the volumes of gaseous reactants and products are in small whole number ratios.
   B. Avogadro's hypothesis: Equal volumes of gases (at the same temperature and pressure) contain the same number of molecules.
   C. Avogadro suggested that certain gases contain diatomic rather than monatomic molecules.

6.3. Avogadro's Number
  A. 6.02×10^{23} is the number of atoms in the atomic weight in grams of an element.
    1. S: atm. wt. = 32.0 = 6.02×10^{23} atoms of S.

6.4. The Mole: "A dozen eggs and a mole of sugar, please."
  A. Mole is a term that refers to 6.02×10^{23} things. It is the amount of substance containing as many elementary units as there are atoms in exactly 12 g of carbon-12 isotope.
    1. One mole of Mg atoms = 6.02×10^{23} atoms.
    2. One mole of H_2O molecules = 6.02×10^{23} molecules.
  B. Molar mass or gram formula weight: mass of 1 mole of substance.
  C. Molar volume: 22.4 L at STP
    1. Molar volume is the volume occupied by one mole of any gas if measured at STP (standard temperature and pressure—0°C and 1 atm). The molar volume equals 22.4 L.
      a. Use the molar volume (22.4 L) to calculate density for any gas for which the molecular formula is known.
  D. Calculations: grams to moles and moles to grams.
    1. The key to these problems is determining a conversion ratio between moles and grams of a substance.
    2. Use the periodic table to calculate the molar mass of a substance. This yields the number of grams in 1 mole.
      a. This ratio can then be used in either type of conversion problem (moles/grams or grams/moles).

6.5 Mole and Mass Relationships in Chemical Equations
  A. Molar masses of substances give the mole-to-gram ratio of a substance.
  B. Coefficients in balanced equations provide mole ratios of the substances in the equation.
  C. Stoichiometry: the mass relationships in chemical equations.
    1. Typically the amount of a substance in a problem is given in grams rather than moles, but when you are asked for grams:
      a. Write a balanced equation.
      b. Use the periodic table to determine the molar masses of substances involved in the problem.
      c. Convert the quantity given to moles by using its molar mass.
      d. Use the mole ratio from the coefficients in the balanced equation to convert moles of the substance given to moles of the substance asked for.

6.6. The Gas Laws
  A. Kinetic-molecular theory.
    1. Matter is composed of tiny, discrete particles.
    2. Particles, which are in constant, random motion move in straight lines.
    3. Particles of a gas are small compared to the distances between them.
    4. Little attraction exists between particles of a gas.
    5. When particles collide, energy is transferred between them. No energy is lost.
    6. Temperature is a measure of the average kinetic energy of gas particles.
  B. Chemists use the kinetic-molecular theory to explain the properties of solids, liquids, and gases.
    1. Solids have an ordered array of particles that are close together.
    2. Liquids have particles that are free to move about, but are close together.
    3. Gases have particles that move at random and are far apart.
  C. Boyle's Law: For a given amount of gas at constant temperature, the volume is inversely proportional to pressure.
  D. Charles's Law: For a given amount of gas at constant pressure, the volume is proportional to absolute temperature.
    1. When the volume of gas reaches zero, the temperature is at absolute zero.

E. The Ideal Gas Law: When Boyle's Law and Charles's Law are combined, the equation PV = nRT is obtained. R is the Universal Gas Constant with a value of $0.0821 \frac{L \cdot atm}{mol \cdot k}$

6.7 Solutions
A. Solutions are homogeneous mixtures with the dissolving substance called the solvent and substance being dissolved called the solute. Molarity is abbreviated as M and has the units of Moles/Liter.
B. Solution Concentrations
1. Molarity is defined as the number of solute moles divided by liters of solution.
a. 5 moles of salt dissolved in 15 liters of solution has a molarity of 3 M.
b. To find the molarity of 54.5 g of KBr dissolved in 3 liters of solution, first convert the grams of KBr to molarity, then divide by the volume.
One mole of KBr = 119.0 g
54.5 g KBr = 0.5 moles
0.5 moles of KBr/3 liters = 0.17 moles/liter or 0.16 M
c. To find the number of grams in a given volume of solution, calculate the number of moles from molarity: M = moles/liters and convert to grams.
How many grams of LiCl are in 300 mL of 0.4 M LiCl?
300 mL is 0.3 L
(0.4 M LiCl) x (0.3 L) = 0.12 moles
LiCl has a molecular mass of 42.5 g/moles
(0.12 moles) x (42.5 g/moles) = 5.1 g LiCl
C. Percent
1. Percent is a fraction (or part/whole) times 100.
a. Divide the mass (volume) of solute by the mass (volume) of solution. Note the solution is the sum of both the solvent and solute.
b. What is the percent $MgCl_2$ in a solution of 15 g $MgCl_2$ in 250 g of solution?
% $MgCl_2$ = 15 g $MgCl_2$/250 g solution x 100
% $MgCl_2$ = 6%

CHAPTER OBJECTIVES

(You should be able to...)

1. Define or identify each of the key terms in this chapter.

2. Given the name, write the symbol for each ion in Table 6.1 of the text, or given the symbol, write the name of the ion.

3. Given the name, write the formula and charge for each polyatomic ion in Table 6.2 of the text, or given the formula and charge, write the name of the ion.

4. Write balanced formulas by combining ions of opposite charge and name the compounds using the ion in Table 6.1 and Table 6.2.

5. Given the name, write balanced formulas for simple covalent compounds, or given the formula, write the name.

6. Given the balanced formula, calculate the molar masses of compounds.

7. Given the balanced formula and a given mass of a substance, calculate the number of moles, or given the number of moles and a balanced formula, calculate the mass of a sample.

8. Be able to use the KMT to explain the properties of solids, liquids, and gases.

9. Use Boyle's Law to calculate V given T or T given V.

10. Use Charles's Law to calculate P given T or T given P.

11. Use Ideal Gas Law to calculate P, V, N, or T given the other variables.

12. Balance simple chemical equations.

13. Use equations to calculate mole and mole ratios of reactants and products.

DISCUSSION

Chapter 5 introduces the language of chemistry. If, instead of chemistry, English literature were our area of study, we would just be at the point of having learned to read. In earlier chapters, we learned to use chemical symbols to represent elements. The symbols are the alphabet of chemistry. In this chapter, we learn to use the "words" of chemistry—the formulas for compounds to compose chemical "sentences"—that is, to write chemical equations. This chapter also introduces the math of chemistry.

You may never have realized how much information is contained in a chemical equation. Much of the chapter is devoted to a discussion of the concepts and terminology needed to extract from an equation every last bit of information. Therefore, one of the first things you should do is to make sure that you understand the Key Terms.

Molar mass and mole are interrelated terms. The molar mass of a compound expressed in grams (the gram formula weight) is the weight of one mole of the compound.

$$1 \text{ gram formula weight} = 1 \text{ mole}$$

You can treat this relationship as a conversion factor. If you calculate the formula weight of a compound and remember the relationship, you can interconvert units expressed in grams to moles or molecules and vice versa.

Example. Interconversions involving methane (CH_4).

Formula weight:

> atomic weight of C = 12   1 x 12 = 12
> atomic weight of H = 1   4 x 1 = 4
> molar mass   = 16

Once we have the molar mass expressed in grams (gram formula weight), we know that there are 16 g in one mole and this becomes our conversion ratio

How many moles are in 4 g of CH_4?

$$16 \text{ g } CH_4 = 1 \text{ mole } CH_4$$

$$4 \text{ g } CH_4 = 0.25 \text{ mole } CH_4$$

How many grams are in 4 moles of CH_4?

$$1 \text{ mole } CH_4 = 16 \text{ g } CH_4$$

$$4 \text{ moles } CH_4 = 64 \text{ g } CH_4$$

Problems 65–76 at the end of Chapter 6 in the text will give you additional practice in this area.

Boyle's Law, Charles's Law, and the Ideal Gas Law quantitatively describe the behavior of gases. Here are some worked examples:

The pressure of a 4.0 L volume of He gas is increased from 5 atmospheres to 7 atmospheres. What is the new volume?

> $P_1V_1 = P_2V_2$
>
> $V_2 = P_1V_1/P_2$
>
> $V_2 = (5 \text{ atm})(4 \text{ L}) /7 \text{ atm}$
>
> $V_2 = 2.9 \text{ L}$

The temperature of a 4.0 L volume of He gas is increased from 10°C to 100°C. What is the new volume?

> $10°C = 283K, 100°C = 373K$
>
> $V_1/T_1 = V_2/T_2$
>
> $V_2 = T_2V_1/T_1$
>
> $V_2 = (323 \text{ K})(4.0 \text{ L})/283 \text{ K}$
>
> $V_2 = 4.6 \text{ L}$

Use the Ideal Gas Law to calculate the volume of 3 mol of He gas at 473K and 2 atmospheres.

$$PV = nRT$$

$$V = nRT/P$$

F. $\quad V = (3 \text{ mol})(0.0821 \underline{\text{ L} \cdot \text{atm}}) (473K) / 2 \text{ atm}$
$$\text{mol} \cdot \text{k}$$

$$V = 58.2 \text{ L}$$

ADDITIONAL PROBLEMS

1. Calculate the gram formula weights of the following compounds. We're using relatively complicated formulas just to make sure that you understand when a subscript applies to a particular atom in the formula and when it does not. You'll require a periodic table or a list of atomic weights.

   a. $CaCO_3$ b. $(NH_4)_2CO_3$ c. $Be(NO_3)_2$ d. $(NH_4)_2C_2O_4$

   e. $Al_2(C_2O_4)_3$ f. $Ca(C_2H_3O_2)_2$ g. $(CH_3)_2SO_4$

2. In this problem, all questions refer to the compound $C_5H_8O_2$.

   a. How many moles of $C_5H_8O_2$ are in 100 g? in 200 g? in 25 g? in 3.687 g?

   b. How many grams of $C_5H_8O_2$ are in 1 mole of the compound? in 8 moles? in 0.8 mole? in 0.01 mole?

3. For additional practice in interconverting these units, answer the questions in problem 2 for the compound H_2CO_3. A calculator would be useful because these numbers will not be easy.

   Let's suppose you are now totally at ease with moles and formula weights and such. That brings us to equations. The first thing you should check in an equation is whether it is balanced or not. (We'll supply the correct reactants and products.) If the equation isn't balanced, the quantitative information derived from it may be incorrect. As we indicated in the chapter, you won't be balancing extremely complex equations, but you should be able to handle those shown in problems 51–54 at the end of the text Chapter 6. Here are some additional equations to balance.

4. Balance the following equations. All formulas are already balanced using subscripts.

   a. $Zn + KOH \rightarrow K_2ZnO_2 + H_2$
   b. $HF + Si \rightarrow SiF_4 + H_2$
   c. $B_2O_3 + H_2O \rightarrow H_6B_4O_9$
   d. $SiCl_4 + H_2O \rightarrow SiO_2 + HCl$
   e. $SnO_2 + C \rightarrow Sn + CO$
   f. $Fe_2O_3 + CO \rightarrow FeO + CO_2$
   g. $Fe_3O_4 + C \rightarrow Fe + CO$
   h. $Fe(OH)_3 + H_2S \rightarrow Fe_2S_3 + H_2O$
   i. $Ca_3P_2 + H_2O \rightarrow PH_3 + Ca(OH)_2$
   j. $Bi_2O_3 + C \rightarrow Bi + CO$

Once you have a balanced equation the coefficients in that equation give you the following information directly.

  a. The combining ratio of molecules (or other formula units such as ion pairs)
  b. The combining ratio of moles of molecules (or other formula units).

The coefficients <u>do not</u> directly give you the combining mass ratios. Thus, from the equation

$$CH_4 + 2\,O_2 \rightarrow CO_2 + 2\,H_2O$$

you know that

  a. 1 molecule of methane (CH_4) reacts with 2 molecules of oxygen (O_2) to give 1 molecule of carbon dioxide (CO_2) and 2 molecules of water (H_2O).
  b. 1 mole of methane reacts with 2 moles of oxygen to give 1 mole of carbon dioxide and 2 moles of water.

The equation does not say that 1 gram of methane reacts with 2 grams of oxygen to produce 1 gram of carbon dioxide and 2 grams of water. If you want to find out how many grams of oxygen react with 1 gram of methane, you must first convert oxygen grams to moles and only then use the chemical equation to determine the combining ratio in moles. After using the equation, you'll have the answer in moles and must then convert to grams.

The examples in the text demonstrate the use of equations to obtain information about combining ratios. Review those examples and then the problems at the end of the chapter. For more practice, try the following problems.

5. Refer to the equation

$$CS_2 + 2\,CaO \rightarrow CO_2 + 2\,CaS$$

  a. How many moles of CO_2 are obtained from the reaction of 2 moles of CS_2? from the reactions of 2 moles of CaO?
  b. How many moles of CaO are consumed if 0.3 mole of CS_2 reacts? if 0.3 mole of CaS is produced?
  c. How many grams of CaS are obtained if 152 g of CS_2 is consumed in the reaction? if 7.6 g of CS_2 is consumed? if 22 g of CO_2 is produced? if 44 g of CO_2 is produced?
  d. How many grams of CaO are required to react completely with 38 g of CS_2? with 152 g of CS_2? to produce 36 g of CaS?

SELF-TEST

Multiple Choice

1. One mole of sulfuric acid, H_2SO_4, has a mass of

  a. 7 g   b. 50 g
  c. 98 g   d. 100 g

2. 3 moles of methane, CH_4, has a mass of

    a. 16 g   b. 32 g
    c. 48 g   d. 64 g

3. NaCl has the gram formula weight (mass) of

    a. 2 g   b. 28 g
    c. 59 g   d. 118 g

4. 34 g of NH_3 ammonia, is _____ moles

    a. 1   b. 2
    c. 3   d. 4

5. How many moles of methane, CH_4, are contained in 48 g?

    a. 1   b. 2
    c. 3   d. 4

6. How many atoms are there in 2 moles of helium?

    a. 2×10^{23} b. 6.02×10^{23}
    c. 12.04×10^{23} d. 6.02×10^{46}

7. How many moles are there in 12×10^{23} molecules of ammonia, NH_3?

    a. 1   b. 2
    c. 3   d. 4

8. One-half mole of SO_2 has a mass of

    a. 8 g   b. 12 g
    c. 24 g   d. 32 g

9. Which is not part of the kinetic molecular theory?

    a. matter is composed of small particles
    b. particles of matter are always in motion
    c. as particles collide, energy is lost to heat
    d. temperature measures the average kinetic energy of the particles

10. Calculate the new volume of a 50 mL volume of a gas at 2 atm when the pressure is increased to 8 atm.

    a. 50 mL   b. 25 mL
    c. 12.5 mL   d. 6.75 mL

11. Calculate the new pressure of a 100 mL volume of a gas at 3 atm when the pressure drops to 5 atm.

    a. 100 atm   b. 30 atm
    c. 60 atm   d. 90 atm

12. Calculate the new volume of a 200 mL volume of a gas when it is heated from 100°C to 200°C.

    a. 100 mL   b. 300 mL
    c. 158 mL   d. 254 mL

13. Calculate the new temperature of a 300 mL volume of a gas at 50°C when expanded to 400 mL.

    a. 67 K   b. 38 K
    c. 431 K   d. 242 K

14. Use the gas law to calculate the volume of 3 moles of O_2 at 2 atm and 150°C.

    a. 104 L   b. 18.5 L
    c. 52 L   d. 37 L

15. Given: $N_2 + 3 H_2 \rightarrow 2 NH_3$. If 2 volumes of N_2 react to form 4 volumes of NH_3, how many volumes of H_2 are needed to form 4 volumes of NH_3?

    a. 1   b. 2
    c. 4   d. 6

16. Given: $CH_4 + 2 O_2 \rightarrow CO_2 + 2 H_2O$. If 1 volume of CH_4 needs 2 volumes of O_2, how many volumes do 3 volumes of CH_4 need?

    a. 1   b. 2
    c. 3   d. 6

17. If 1 volume of ethane, C_2H_6, contains 6.02×10^{23} molecules, then 3 volumes contain

    a. 6×10^{23} b. 12×10^{23}
    c. 18×10^{23} d. 24×10^{23}

18. At STP, 1 volume of a gas occupies 22.4 L, while 4 volumes occupy:

    a. 22.4 L   b. 44.8 L
    c. 67.2 L   d. 89.6 L

Given the <u>unbalanced</u> chemical equations, what is the coefficient of the underlined chemical in a balanced equation?

19. $N_2 + H_2 \rightarrow \underline{NH_3}$

    a. 1   b. 2
    c. 3   d. 4

20. $C_2H_4 + O_2 \rightarrow \underline{CO_2} + H_2O$

    a. 1   b. 2
    c. 3   d. 4

21. $H_3PO_4 + \underline{Ba(OH)_2} \rightarrow Ba_3PO_4 + H_2O$

    a. 1   b. 2
    c. 3   d. 4

22. What is the molarity of a solution containing 7 moles of solute in 9 liters of solution?

      a. 16 M   b. 63 M

      c. .78 M   d. 1.3 M

23. Calculate the molarity of 53 g. of $LiNO_3$ dissolved in 700 mL of water.

      a. 1.9 M   b. 1.1 M

      c. 0.19 M   d. 0.11 M

24. How many grams of $CaCl_2$ are needed to make a 350 mL of 0.3 M solution?

      a. 11.6 g   b. 1.16 g

      c. 9.5 g   d. 7.4 g

25. Calculate the percent of sugar in a 300 g solution containing 25 g of sugar.

      a. 12%   b. 7.5%

      c. 8.3%   d. 83.3%

ANSWERS

Additional Problems

1.

    a. 100   b. 96   c. 133   d. 124

    e. 318   f. 158   g. 126

2. The formula weight of $C_5H_8O_2 = 100$.

    a. 1 mole; 2 moles; 0.25 mole; 0.03687 mole

    b. 100 g; 800 g; 80 g; 1 g

3. The formula weight of $H_2CO_3 = 62$.

    a. 1.6 moles; 3.2 moles; 0.40 mole; 0.059 mole

    b. 62 g; 496 g; 49.6 g; 0.62 g

4.

    a. $Zn + 2\ KOH \rightarrow K_2ZnO_2 + H_2$

    b. $4\ HF + Si \rightarrow SiF_4 + 2\ H_2$

    c. $2\ B_2O_3 + 3\ H_2O \rightarrow H_6B_4O_9$

    d. $SiCl_4 + 2\ H_2O \rightarrow SiO_2 + 4\ HCl$

    e. $SnO_2 + 2\ C \rightarrow Sn + 2\ CO$

    f. $Fe_2O_3 + CO \rightarrow 2\ FeO + CO_2$

    g. $Fe_3O_4 + 4\ C \rightarrow 3\ Fe + 4\ CO$

    h. $2\ Fe(OH)_3 + 3\ H_2S \rightarrow Fe_2S_3 + 6\ H_2O$

    i. $Ca_3P_2 + 6\ H_2O \rightarrow 2\ PH_3 + 3\ Ca(OH)_2$

    j. $Bi_2O_3 + 3\ C \rightarrow 2\ Bi + 3\ CO$

5.

    a. 2 moles of CO_2; 1 mole of CO_2

    b. 0.6 mole of CaO; 0.3 mole of CaO

    c. 288 g of CaS; 14.4 g of CaS; 72 g of CaS; 144 g of CaS

    d. 56 g of CaO; 224 g of CaO; 28 g of CaO

Self-Test

1. c	8. d	15. d	22. c
2. c	9. c	16. d	23. b
3. c	10. c	17. c	24. a
4. b	11. c	18. d	25. c
5. c	12. c	19. b	
6. c	13. c	20. b	
7. b	14. c	21. c	

CHAPTER 7

Acids and Bases

Please Pass the Protons

KEY TERMS

acidic anhydride	basic anhydride	strong acid
acids	neutralization	strong base
alkalosis	pH	weak acid
base	salt	weak base

CHAPTER SUMMARY

7.1 Acids and Bases: Experimental Definitions
    A. Characteristic properties of acids
        1. Turn litmus red.
        2. Taste sour.
        3. React with active metals to produce H_2 gas.
        4. React with bases to produce water and ionic compounds called salts.
    B. Characteristic properties of bases
        1. Turn litmus blue.
        2. Taste bitter.
        3. Feel slippery.
        4. React with acids to form water and a salt.

7.2 Acids Explained: Hydronium Ions
    A. In water solution, the properties of acids are due to the hydronium ion (H_3O^+).
        1. A hydronium ion is a proton (H^+) attached to a water molecule.
        2. In acid-base chemistry, the proton is the same as the hydrogen ion (H^+).
    B. In general, acids are defined as proton donors. (Arrhenius definition as well as Bronsted-Lowry definition of acids)
    C. In water solution, the properties of bases are due to the hydroxide ion (OH^-). (Arrhenius definition of bases)
    D. When ionic compounds such as NaOH and $Ca(OH)_2$ are dissolved in water, they dissociate into separate ions. The hydroxide ions make the solution basic.
    E. Ammonia is basic because it reacts with water to produce hydroxide ions.
$$NH_3 + H_2O \rightarrow NH_4^+ + OH^-$$
    F. In general, bases are proton acceptors. (Bronsted-Lowry definition of bases)

7.3   Acidic and Basic Anhydrides
    A. Nonmetal oxides are called acidic anhydrides. They form acids when added to water.
    B. Metal oxides are called basic anhydrides. They react with water to form bases.

7.4. Strong and Weak Acids and Bases
    A. Strong acids are those that ionize completely, or nearly so, in aqueous solutions.
$$HCl + H_2O \rightarrow H_3O^+ + Cl^-$$
    B. Weak acids ionize only slightly in aqueous solutions. (A mole of HCN in 1 L of water solution is only 1.0 % dissociated.)
$$HCN + H_2O \rightarrow H_3O^+ + CN^-$$
    C. Strong bases are those that dissociate completely, or nearly so, in water.
    D. Weak bases are those that yield relatively few hydroxide ions when dissolved in water.
    E. Neutralization.
        1. In water, an acid will release H_3O^+ and when combined with an equivalent amount of base that releases OH^-, a neutralization reaction combining H_3O^+ (acid) and OH^- (base) will produce water and a salt.
$$HCl + NaOH \rightarrow H_2O + NaCl$$
$$Acid + Base \rightarrow Water + Salt$$
$$H_3O^+ + OH^- \rightarrow H_2O$$

7.5   The pH Scale
    A. pH is a measure of the concentration of H_3O^+ in mole/liter.
        1. On the pH scale, a concentration of 1×10^{-7} mol/L of hydronium ions becomes a pH of 7; a concentration of 1×10^{-10} mol/L becomes a pH of 10; and so on.
$$pH = -\log [H_3O^+]$$
        2. Interpreting pH.
            a. pH < 7 is acidic
            b. pH = 7 is neutral
            c. pH > 7 is basic
    B. pOH is a measure of the concentration of OH^- in moles/liter.
        1. pOH is related to pH by the following equation:
$$pH + pOH = 14$$

7.6   Acid Rain
    A. All rain is slightly acidic because of atmospheric CO_2.
    B. "Acid rain" is produced when acidic pollutants such as sulfur oxides and nitrogen oxides are present in the atmosphere.

7.7   Antacids: A Basic Remedy
    A. The stomach secretes HCl to aid in digestion of food. Overindulgence and stress can lead to hyperacidity and ulcers.
    B. Antacids are bases.
        1. Sodium bicarbonate ($NaHCO_3$), also known as baking soda, can neutralize an upset stomach.
            a. Overuse of sodium bicarbonate can lead to alkalosis, a condition whereby the blood becomes too basic.
        2. Calcium carbonate ($CaCO_3$) is found in Tums.
            a. Regular use can lead to constipation.

7.8   Acid and Bases in Industry and in the United States.
    A. Strong acids and bases are corrosive poisons that can cause chemical burns.

B. Most acids and bases produced in this country are used by industry, but some can be found around the house.
   1. H_2SO_4 is in automobile batteries, some drain cleaners.
   2. HCl, hydrochloric acid or commonly called muriatic acid is used to remove rust.
   3. CaO (lime) is used in cement, mortar, plaster, and on lawns.
   4. NaOH (lye) is used in drain and oven cleaner.
C. Acids and bases in health and disease
   1. Strong acids and bases are corrosive (break down protein molecules).
   2. The proper pH is critical to many biochemical reactions in the body.

CHAPTER OBJECTIVES

(You should know that...)

1. Acidic solutions taste sour, turn litmus red, neutralize bases, and dissolve active metals to release hydrogen gas (H_2).

2. In water H^+ ions are combined with H_2O to produce H_3O^+ ions.

3. The properties of aqueous acids are due to hydronium ions (H_3O^+).

4. Basic solutions taste bitter, feel soapy, turn litmus blue, and neutralize acids.

5. The properties of aqueous bases are due to hydroxide ions (OH^-).

6. Vinegar, citrus juices, and stomach fluid are acidic solutions.

7. Household ammonia, solid drain cleaners, and oven-cleaner preparations contain basic compounds.

8. In a general sense, acids are hydronium ion (H_3O^+) donors.

9. In a general sense, bases are proton (H^+) acceptors.

10. Strong acids react completely with water, yielding a lot of hydronium ions.

11. Weak acids yield relatively few hydronium ions in water solution.

12. Hydrochloric acid (HCl) is "stomach acid."

13. Nonmetal oxides (acidic anhydrides) react with water to form acids. Metal oxides (basic anhydrides) react with water to form bases. You should also be able to write the equation for both types of reactions.

14. All antacids contain a basic compound.

15. Both strong acids and bases can cause chemical burns.

16. A pH value less than 7 is acidic, a pH value greater than 7 is basic, and a pH value of 7 is neutral.

17. Sulfuric acid is the leading chemical product of U.S. industry.

18. Acid rain is caused by acidic pollutants, sulfur oxides, and nitrogen oxides.

(You should be able to...)

19. Match the formulas and names of the acids and bases used in the text.

20. Classify acids and bases as strong or weak depending on the relative amount of ions they release.

21. Determine the pH of a solution given the hydronium ion in exponential form and determine the hydronium ion concentration of a solution, given the pH (whole numbers only).

DISCUSSION

We will begin by reviewing the definitions of acids. Acids are defined in the chapter in two ways: as compounds that yield hydronium ions in aqueous solutions and as compounds that act as proton donors. It is now generally accepted that a hydrogen ion does not exist in solution as an independent unit. Thus, the second definition is an attempt to be a bit more accurate in describing the action of an acid.

Since the H^+ ion is only a proton (a nuclear particle), it exists in water only associated with H_2O as the H_3O^+ ion called the hydronium ion. There are many practical aspects of acids and bases that you should learn: their properties, their sources (acid and base anhydrides), whether strong or weak, the pH scale, and their applications. Practice writing formulas and equations for each of these.

SELF-TEST

Matching
Match the formulas of these acids and bases with the proper name.

___1. H_2SO_4 a. sodium hydroxide
___2. HCl   b. ammonia
___3. NaOH   c. hydrocyanic acid
___4. HCN   d. potassium hydroxide
___5. KOH   e. sulfuric acid
___6. $Ca(OH)_2$ f. calcium hydroxide
___7. NH_3 g. hydrochloric acid

Multiple Choice

8.   In aqueous solution, a substance is judged to be an acid if it produces an excess of

        a. hydronium ions   b. electrons
        c. anions   d. hydroxide ions

9.   In aqueous solution, a substance is judged to be a base if it produces an excess of

        a. hydronium ions   b. protons
        c. cations   d. hydroxide ion

10. Which is not characteristic of acids?

    a. taste sour
    b. neutralize bases
    c. turn litmus paper blue
    d. react with active metals to yield H_2 gas

11. If litmus paper changes from red to blue when placed in an aqueous solution, the solution is

    a. too hot   b. acidic
    c. basic   d. neutral

12. All the following are bases except

    a. KOH   b. NaOH
    c. NH_3 d. HCN

13. All the following are acids except

    a. HCl   b. HCN
    c. H_2SO_4 d. NH_3

14. What product is formed: $SO_3 + H_2O \rightarrow$

    a. HSO   b. HSO_2
    c. H_2SO_3 d. H_2SO_4

15. Which is a weak base in aqueous solution?

    a. NH_3 b. KOH
    c. NaOH   d. $Ca(OH)_2$

16. What product is formed $MgO + H_2O \rightarrow$

    a. MgHO   b. H_2MgO_2
    c. $Mg(OH)_2$ d. MgO_2H_2

17. HCN reacts only slightly with water, producing relatively few hydronium ions. Therefore, HCN is a

    a. weak acid   b. strong acid
    c. weak base   d. strong base

18. A solution of pH 7 is

    a. acidic   b. basic
    c. exactly neutral   d. salty

19. The effluent from a factory dissolves iron metal, turns litmus paper red, and tastes sour. The effluent probably contains

    a. hydrochloric acid   b. hydroxide ions
    c. salt   d. alcohol (C_2H_5OH)

20. Which is a weak acid?

    a. HCl   b. HNO_3
    c. H_2SO_4 d. HCN

21. Which is the strongest acid?

    a. HCl   b. CH_3COOH
    c. HCN   d. H_3BO_3

22. Which is strong base in aqueous solution?

    a. NH_3 b. HCN
    c. NaOH   d. $NaHCO_3$

23. A pH of 5 has a hydronium ion concentration of

    a. 0.001 M   b. 0.0001 M
    c. 0.00001M   d. 0.000001 M

24. Which pH represents the most acidic solution?

    a. 6   b. 1
    c. 7   d. 13

25. 0.001 M HCl has a pH of

    a. 1   b. 2
    c. 3   d. 4

26. Which is <u>not</u> an acidic solution?

    a. lemon juice   b. orange pop
    c. household ammonia   d. stomach fluid

27. A pH of 11 has a hydroxide concentration of

    a. 0.1 M   b. 0.01 M
    c. 0.001 M   d. 0.0001 M

28. The acid found in stomach fluid is

    a. NH_3 b. CH_3COOH
    c. HCl   d. H_2SO_4

29. 0.01 M NaOH has a pH of

    a. 14   b. 13
    c. 12   d. 11

30. A solution of pH 5 is

    a. acidic   b. basic
    c. exactly neutral   d. bitter to the taste

31. All antacids

    a. are proton donors
    b. contain hydrochloric acid (HCl)
    c. are basic compounds
    d. contain hydroxide ions

32. Which of the following is responsible for the properties of all aqueous bases? (Aqueous means in water.)

    a. H_3O^+ b. OH^-
    c. H_2O_2 d. Na^+

33. In water, hydrogen chloride forms acidic solutions because it reacts with water to form

    a. H_2 b. H_3O^+
    c. OH^- d. ClO^-

34. A solution of ammonium nitrate in water has a pH of 2. The solution is

    a. strongly acidic   b. strongly basic
    c. weakly acidic   d. weakly basic

35. Which pH corresponds to a mildly basic solution?

    a. 1   b. 5
    c. 9   d. 14

36. The leading product of the U.S. chemical industry is

    a. NH_3 b. NaOH
    c. H_2SO_4 d. DDT

37. In the reaction $CH_3NH_2 + H_2O \rightarrow CH_3NH_3^+ + OH^-$ the compound CH_3NH_2 is a(n)

    a. acid   b. base
    c. oxidizing agent   d. reducing agent

38. In the reaction $C_6H_5OH + H_2O \rightarrow C_6H_5O^- + H_3O^+$ the compound C_6H_5OH is a(n)

    a. acid   b. base
    c. oxidizing agent   d. reducing agent

39. Acid rain is caused by

    a. CO_2 b. SO_2
    c. CH_4 d. NH_3

40. Normally rain is slightly acidic because of dissolved

    a. CO_2 b. SO_2
    c. CH_4 d. NH_3

ANSWERS

1. e	9. d	17. a	25. c	33. b
2. g	10. c	18. c	26. c	34. a
3. a	11. c	19. a	27. c	35. c
4. c	12. d	20. d	28. c	36. c
5. d	13. d	21. a	29. c	37. b
6. f	14. d	22. c	30. a	38. a
7. b	15. a	23. c	31. c	39. b
8. a	16. c	24. b	32. b	40. a

Oxidation and Reduction

Burn and Unburn

KEY TERMS

activation energy

anode

antioxidant

battery

bleach

catalyst

cathode

electrochemical cell

fuel cell

oxidation

oxidizing agent

photosynthesis

reducing agent

reduction

CHAPTER SUMMARY

8.1   Oxidation and Reduction: Three Definitions

   A. Oxidation is the <u>gain</u> of oxygen atoms.
        Reduction is the <u>loss</u> of oxygen atoms

      1. $4\,Fe + 3\,O_2 \rightarrow 2\,Fe_2O_3$
            Fe is oxidized.
            O_2 is reduced.

      2. $2\,H_2 + O_2 \rightarrow 2\,H_2O$
            H_2 is oxidized.
            O_2 is reduced.

   B. Oxidation is a <u>loss</u> of hydrogen atoms.
        Reduction is a <u>gain</u> of hydrogen atoms.

      1. $CH_4O + 1/2\,O_2 \rightarrow CH_2O + H_2O$
            CH_4O is oxidized.

      2. $CO + 2\,H_2 \rightarrow CH_4O$
            CO is reduced.

   C. Oxidation is a <u>loss</u> of electrons.
        Reduction is a <u>gain</u> of electrons.

      1. $Mg + Cl_2 \rightarrow Mg^{2+} + 2\,Cl^-$
            Mg is oxidized.
            Cl_2 is reduced.

   D. <u>Increase</u> in oxidation state is <u>oxidation</u>
        <u>Decrease</u> in oxidation state is <u>reduction</u>.

      1. $Mg + \frac{1}{2}\,O_2 \rightarrow Mg^{+2}O^{-2}$.
            Mg is oxidized.
            O is reduced.

8.2 Oxidizing and Reducing Agents.
  A. Oxidation and reduction must occur together.
    1. Substance being <u>oxidized</u> is the <u>reducing agent</u>.
    2. Substance being <u>reduced</u> is the <u>oxidizing agent</u>.

8.3 Electrochemistry: Cells and Batteries
  A. Electricity can cause chemical change (electrolysis). Chemical change can produce electricity (batteries).
  B. When a reactive metal (zinc) is placed in contact with the ions of a less reactive metal (copper), the more active metal will give up its electrons (oxidation) to the ions of the less active metal (which are reduced).
  C. If the two metals (zinc and copper) are placed in solutions of their metal ions in separate containers, the electrons traveling between the two containers must flow through an external circuit and can be harassed to do work.
    1. The metal electrode where oxidation takes place (more active metal) is the <u>anode</u>.
    2. The metal electrode where reduction takes place (less active metal) is the <u>cathode</u>.
    3. This arrangement is an <u>electrochemical cell</u>.
  D. A battery is a series of electrochemical cells. (In everyday life, however, we refer to a single electrochemical cell, such as that used in flashlights, as a "battery.")
  E. Dry cells are the common batteries used in flashlights.
  F. Lead storage batteries are the rechargeable batteries found in cars.
  G. Other common batteries are lithium iodine cells used in pace-makers, lithium–FeS_2 batteries used in cameras, rechargeable Ni-Cad battery for portable radios, and small "button" batteries.
  H. In fuel cells, fuel is oxidized at the anode and oxygen is reduced at the cathode.
    1. Fuel cells are a much more efficient way of using fuel.

8.4 Corrosion
  A. The rusting of iron is an electrochemical process that requires water, oxygen, and an electrolyte.
    1. Oxidation and reduction often occur at different places on the metal's surface.
  B. Aluminum is more reactive than iron, but is protected by an aluminum oxide film on its surface that forms from the reaction of Al with oxygen in the air.
  C. Silver tarnish is largely silver sulfide.
    1. It is formed by the reaction of silver with hydrogen sulfide (H_2S) in the air.
      a. Tarnish can be removed by reacting it with aluminum.

8.5 Explosive Reactions: Most explosive reactions are oxidation-reduction reactions such as the ammonium nitrate-fuel oil bomb in Oklahoma City, OK.

8.6 Oxygen: An Abundant and Essential Oxidizing Agent
  A. Oxygen is the most abundant element on this planet.
    1. Air: 21% uncombined oxygen by volume.
    2. Water: 89% by weight combined oxygen.
    3. People: approximately 60% by weight combined oxygen.
  B. Fuels such as natural gas, gasoline, coal, and the foods we eat all need oxygen for combustion to release their stored chemical energy.
  C. Pure oxygen is obtained by liquefying air, then allowing the nitrogen and argon to boil off.
  D. Many metals and nonmetals react with oxygen.
  E. Ozone is a powerful oxidizing agent which is a pollutant at ground level and a beneficial chemical in the upper stratosphere.

8.7 Other Common Oxidizing Agents
   A. Hydrogen peroxide, used as a 3% or 30% aqueous solution, is converted to water during oxidation.
   B. Potassium dichromate is a common laboratory oxidizing agent.
      1. Breathalyzer tests make use of the color change associated with the reduction of dichromate ions.
   C. Many antiseptics are mild oxidizing agents.
   D. Oxidizing agents are also used as disinfectants.
      1. Example: use of chlorine in swimming pools.
   E. Bleach is an oxidizing agent that removes unwanted color from materials.

8.8 Some Reducing Agents of Interest
   A. Elemental carbon (coke) is used as a reducing agent for large scale production of metals.
   B. Photographic developing solutions are reducing agents that "fix" silver in photographic film by reducing Ag^+ ions that have been exposed to light to free Ag metal.
   C. Antioxidants such as ascorbic acid (vitamin C) and vitamin E are reducing agents in food chemistry.

8.9 Hydrogen: Lightweight and Reactive
   A. Hydrogen represents only 0.9% of the Earth's crust (by weight), but is the most abundant element in the universe.
      1. Elemental hydrogen (uncombined) is rarely found on Earth.
      2. Combined hydrogen is found in water, natural gas, petroleum products, and all foodstuffs.
   B. Hydrogen gas is also used as a reducing agent for the production of expensive metals.

8.10 Oxidation, Reduction, and Living Things
   A. Reduced compounds represent a form of stored potential energy. The driving force to produce reduced compounds is ultimately derived from the sun in the processes of photosynthesis.
      1. Plants and the animals that feed on plants use the glucose to make other reduced compounds such as carbohydrates.
   B. We produce energy to meet our metabolic and other body needs by releasing the energy stored in these reduced compounds as we oxidize them back to CO_2 and water.

CHAPTER OBJECTIVES

(You should...)

1. Be able to recognize a substance that is oxidized (it gains oxygen, loses hydrogen, or loses electrons).

2. Be able to recognize a substance that is reduced (it loses oxygen, gains hydrogen, or gains electrons).

3. Know that the substance undergoing oxidation is the reducing agent.

4. Know that the substance undergoing reduction is the oxidizing agent.

5. Recognize each of the following as oxidizing agents: hydrogen peroxide (H_2O_2), bleach, chlorine (Cl_2), and oxygen (O_2).

6. Recognize each of the following as reducing agents: hydrogen (H_2), carbon (C), and photographic developer.

7. Know that an electrochemical cell uses chemical change to produce electricity.

8. Know that a battery is a series of electrochemical cells.

9   Know that electricity is a flow of electrons.

10. Know that corrosion of iron requires water, oxygen, and an electrolyte.

11. Know that explosive reactions are usually oxidation-reduction reactions.

12. Know that aluminum is more reactive than iron, but it is protected from corrosion by an impervious film of aluminum oxide.

13. Know that silver tarnish is largely silver sulfide.

14. Know that disinfectants, antiseptics, and bleaches are frequently oxidizing agents.

15. Know that plants reduce CO_2 to form carbohydrates.

16. Know that animals oxidize carbohydrates to obtain energy.

DISCUSSION

It is impossible to overemphasize the importance of oxidation-reduction processes. Think of it this way: You are powered by the energy of sunlight just like artificial satellites. But you can't unfold solar panels like artificial satellites to convert sunlight to electrical energy. You are a chemical factory and not an artificial satellite, solar-powered or otherwise. So somehow you have to tap that solar energy in a chemical way. This is precisely the role of oxidation-reduction reactions in life processes—they plug you into the sun.

You should practice writing the formulas and chemical equations for the oxidation-reduction reactions. Be sure to know the common oxidation and reduction agents and their applications. These applications range from large scale industrial uses to batteries, to household uses and finally the source of energy for living organisms including us.

EXAMPLE PROBLEM

1. Identify the element being oxidized, the element being reduced, the oxidizing agent, and the reducing agent in the following reactions.
    a. $2\,Mg + CO_2 \rightarrow 2\,MgO + C$
        Mg gains oxygen; therefore, it is oxidized. CO_2 loses oxygen; therefore, it is reduced. If Mg is oxidized, CO_2 must be the oxidizing agent. If CO_2 is reduced, Mg must be the reducing agent.

    b. $C_2H_4 + N_2H_2 \rightarrow C_2H_6 + N_2$
        C_2H_4 gains hydrogen; therefore, it is reduced. N_2H_2 gives up hydrogen; therefore, it is oxidized. N_2H_2 is the reducing agent (it reduces C_2H_4), and C_2H_4 is the oxidizing agent (it oxidizes N_2H_2).
    c. $Cu + 2\,Ag^+ \rightarrow Cu^{2+} + 2\,Ag$

Cu loses electrons to become Cu^{2+}; therefore, it is oxidized (and is the reducing agent). Ag+ gains electrons to become Ag; therefore, it is reduced (and is the oxidizing agent).

ADDITIONAL PROBLEM

1. Identify the substance oxidized, the substance reduced, the oxidizing agent, and the reducing agent in each of the following equations.
   a. $CuO + H_2 \rightarrow Cu + H_2O$
   b. $C_2H_4 + 3\ O_2 \rightarrow 2\ CO_2 + 2\ H_2O$
   c. $Fe + Cu^{2+} \rightarrow Fe^{2+} + Cu$
   d. $5\ CO + I_2O_5 \rightarrow I_2 + 5\ CO_2$
   e. $CH_3CHO + H_2O_2 \rightarrow CH_3COOH + H_2O$
   f. $C_6H_{12}O_6 + 6\ O_2 \rightarrow 6\ CO_2 + 6\ H_2O$
   g. $16\ H^+ + 2\ Cr_2O_7^{2-} + 3\ C_2H_5OH \rightarrow 4Cr^{3+} + 3\ C_2H_4O + 11\ H_2O$

SELF-TEST

Multiple Choice

1. Which substance is an oxidizing agent?

   a. NaOH   b. NaOCl
   c. HCl   d. NaCl

2. Which substance is a reducing agent?

   a. H_2 b. F_2
   c. Cl_2 d. I_2

3. When CH_4 is burned, the products are

   a. $C + H_2$ b. $CH_2 + H_2O$
   c. $CO_2 + H_2$ d. $CO_2 + H_2O$

4. When sulfur (S) burns in oxygen (O_2) the product is

   a. SO_2 b. H_2S
   c. H_2SO_4 d. CO_2

5. Hydrogen peroxide converts lead sulfide (PbS) to lead sulfate (PbSO$_4$). The lead sulfide is

   a. a base   b. an oxidizing agent
   c. oxidized   d. reduced

6. At ground level ozone is _____ while in the upper atmosphere it is_____

   a. harmful, harmful   b. beneficial, beneficial
   c. beneficial, harmful   d. harmful, beneficial

7.   In which of the following is the reactant undergoing oxidation? (These are not complete chemical equations.)

    a. $Cl_2 \rightarrow 2\,Cl^-$ b. $WO_3 \rightarrow W$
    c. $2\,H^+ \rightarrow H_2$ d. $CO \rightarrow CO_2$

8.   Which substance is a reducing agent?

    a. HCl   b. F_2
    c. C   d. I_2

9.   Which substance is an oxidizing agent?

    a. NaOH   b. Cl_2
    c. H_2 d. NaCl

10.   When C_3H_8 is burned, the products are

    a. $C + H_2$ b. $CH_2 + H_2O$
    c. $CO_2 + H_2$ d. $CO_2 + H_2O$

11.   Hydrogen gas converts tungsten oxide (WO) to tungsten (W) metal. Hydrogen (H_2) is a(n)

    a. acid   b. base
    c. oxidizing agent   d. reducing agent

12.   In the reaction $H_2CO + H_2O_2 \rightarrow H_2CO_2 + H_2O$ the compound H_2CO (formaldehyde) is

    a. oxidized   b. reduced
    c. an acid   d. a base

13.   In the reaction $MoO_3 + 3\,H_2 \rightarrow Mo + 3\,H_2O$ the molybdenum oxide (MoO_3) is

    a. oxidized   b. reduced
    c. an acid   d. a base

14.   Sodium thiosulfate ($Na_2S_2O_3$) converts iodine (I_2) to iodide ions (I^-). Sodium thiosulfate is a(n)

    a. acid   b. base
    c. oxidizing agent   d. reducing agent

15.   In the reaction $SnO_2 + H_2 \rightarrow SnO + H_2O$ the SnO_2 is

    a. oxidized   b. reduced
    c. an acid   d. a base

16.   In the reaction $2\,I^- + Cl_2 \rightarrow I_2 + 2\,Cl^-$ the chlorine (Cl_2) is

    a. oxidized   b. reduced
    c. both oxidized and reduced   d. neither oxidized nor reduced

17. Photographic developer acts as

    a. an acid   b. a base
    c. an oxidizing agent   d. a reducing agent

18. Batteries are an example of

    a. acid/base reactions   b. oxidation/reduction
    c. combination reaction   d. decomposition reaction

19. In making sugars from carbon dioxide and water, green plants act as

    a. acids   b. bases
    c. oxidizing agents   d. reducing agents

20. In a reaction, the substance undergoing oxidization serves as the

    a. reducing agent   b. oxidizing agent
    c. electron acceptor   d. proton acceptor

21. A substance is oxidized if it

    a. gains oxygen atoms   b. gains hydrogen atoms
    c. gains electrons   d. all of these

22. Disinfectants often are

    a. strong acids   b. strong bases
    c. oxidizing agents   d. reducing agents

23. Electric current is a flow of

    a. gamma rays   b. electrons
    c. neutrons   d. protons

24. In photosynthesis carbon dioxide is _____ to form carbohydrates

    a. acidified   b. made basic
    c. oxidized   d. reduced

25. A device that converts chemical energy to electricity is called a (n)

    a. electrocardiograph   b. electrochemical cell
    c. transformer   d. voltmeter

26. Explosive reactions are usually

    a. acid/base   b. oxidation/reduction
    c. combination   d. decomposition

ANSWERS

Additional Problems

1. a. CuO is reduced; it is the oxidizing agent. H_2 is oxidized; it is the reducing agent.
    b. C_2H_4 is oxidized; it is the reducing agent. O_2 is reduced; it is the oxidizing agent.
    c. Fe is oxidized; it is the reducing agent. Cu^{2+} is reduced; it is the oxidizing agent.
    d. CO is oxidized; it is the reducing agent. I_2O_5 is reduced; it is the oxidizing agent.
    e. CH_3CHO is oxidized; it is the reducing agent. H_2O_2 is reduced; it is the oxidizing agent.
    f. $C_6H_{12}O_6$ is oxidized; it is the reducing agent. O_2 is reduced; it is the oxidizing agent.
    g. $Cr_2O_7^{2-}$ is reduced; it is the oxidizing agent. C_2H_5OH is oxidized; it is the reducing agent.

Self-Test

1. b	7. d	13. b	19. d	25. b
2. a	8. c	14. d	20. a	26. b
3. d	9. b	15. b	21. a	
4. a	10. d	16. b	22. c	
5. c	11. d	17. d	23. b	
6. d	12. a	18. b	24. d	

Organic Chemistry

The Infinite Variety of Carbon Compounds

KEY TERMS

addition reaction
alcohol (ROH)
aldehyde (RCHO)
alkaloid
alkane
alkene
alkyl group (R-)
alkyne
amide
amine
amino group (-NH$_2$)

aromatic compound
carbonyl group (-CO-)
carboxyl group (-COOH)
carboxylic acid (RCOOH)
condensed structural formula
cyclic hydrocarbon
ester (RCOOR')
ether (R-OR')
functional group
heterocyclic compound
homologous series

hydrocarbon
isomers
ketone (R-CO-R')
organic compound
phenol
saturated hydrocarbons
structural formula
unsaturated hydrocarbon

CHAPTER SUMMARY

9.1   The Unique Carbon Atom
   A. Carbon can form long chains of atoms (catenation).
   B. Other unique properties
      1. The chains can form branches and rings.
      2. Carbon can bond to nearly all the other elements, but it forms especially strong bonds to hydrogen, oxygen, and nitrogen.

Hydrocarbons
   The simplest organic compounds contain only carbon and hydrogen.
9.2   Simple Hydrocarbons: Alkanes
   A. Hydrocarbons are compounds consisting of only carbon and hydrogen.
   B. Alkanes are saturated hydrocarbons where each carbon is bonded to four atoms; there are no double or triple bonds.
      1. The simplest alkane is methane (CH$_4$).
   C. Structural formulas show the order in which atoms are attached but do not represent the geometry.
   D. Alkanes form a series with formula C$_n$H$_{2n+2}$. The first three members are
      1. Methane, CH$_4$.
      2. Ethane, C$_2$H$_6$ or CH$_3$CH$_3$.

3. Propane, C_3H_8 or $CH_3CH_2CH_3$.

E. Naming alkanes: All alkanes have the "ane" ending with a standard stem used to indicate the number of carbon atoms. These stems are as follows.

meth = 1	hex = 6
eth = 2	hept = 7
prop = 3	oct = 8
but = 4	non = 9
pent = 5	dec = 10

F. Isomerism
1. Isomers have the same molecular formula, but different structural formulas.

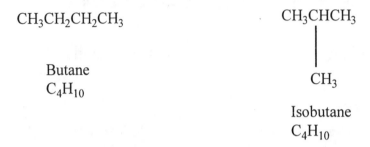

G. Homologous Series: Homology gives organization and meaning to organic chemistry in the same way that the periodic table gives organization and meaning to the chemistry of the elements.
1. In a homologous series (like the alkanes) each member differs from the previous member by a given number of C's and H's.
   a. Alkanes differ from each other by a CH_2 unit.

H. Condensed structural formulas omit almost all bond lines between atoms for simplicity sake. Hydrogens and other groups are written following the carbon atom to which they are attached.

CH_3CH_3 $CH_3CH_2CH_2CH_2CH_3$
Ethane   Pentane

I. Properties of alkanes
1. Alkanes are flammable.
2. Alkanes with 1 to 4 C's are gases at room temperature.
3. Alkanes with 5 to 16 C's are liquids.
4. Alkanes with 17 or more C's are solid.
3. Their densities are lower than that of water.
5. They are nonpolar molecules (insoluble in water).
6. They undergo very few chemical reactions.
7. In the lungs, alkanes can lead to chemical pneumonia.
8. Heavy alkanes (long C chains) act as emollients (skin softeners).

9.3 Cyclic Hydrocarbons: Rings and Things
A. Rings or cyclic structures can be formed with compounds of three or more carbons. These cycloalkanes have properties similar to those of their non-cyclic counterparts.

B. Cycloalkanes are named by using the prefix "cyclo-" with the same stem and suffix as the open-chain compound with the same number of carbon atoms.

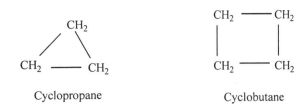

Cyclopropane   Cyclobutane

C. Cyclic structures are represented by geometric figures.
   1. Three carbon: cyclopropane: triangle.
   2. Four carbon: cyclobutane: square.
   3. Five carbon: cyclopentane: pentagon.
   4. Six carbon: cyclohexane: hexagon.

9.4   Unsaturated Hydrocarbons: Alkenes and Alkynes
   A. Alkenes are unsaturated hydrocarbons characterized by a carbon-to-carbon double bond.
      1. Ethylene, $CH_2 = CH_2$ (also called ethene), is the simplest alkene.
      2. Propylene, $CH_3CH = CH_2$ (also called propene) is the next alkene.
   B. Alkynes are unsaturated hydrocarbons characterized by carbon-to-carbon triple bonds.
      1. Acetylene, $HC \equiv CH$ (also called ethyne), is the simplest and most common alkyne.
   C. Alkenes and alkynes are called unsaturated hydrocarbons because they can add more hydrogen atoms (to form saturated hydrocarbons).

$$H_2C \equiv CH_2 + H_2 \longrightarrow CH_3 \text{-} CH_3$$

$$HC \equiv CH + 2H_2 \longrightarrow CH_3 \text{-} CH_3$$

   D. Properties of Alkenes and Alkynes
      1. These compounds undergo addition reactions where hydrogen or halogens are "added" to the double or triple bonds.
      2. Alkenes and alkynes can "add" to each other forming large molecules called polymers.

9.5   Aromatic Hydrocarbons: Benzene and Relatives
   A. Benzene was first isolated in 1825. It has the molecular formula C_6H_6.
   B. In 1865, Kekulé proposed a cyclic structure with alternating single and double carbon-carbon bonds for benzene. However, it has been determined that all of the bonds in benzene are identical and that there are no real double bonds.
      1. Benzene doesn't react by addition reactions the way unsaturated hydrocarbons do, but rather by substitution reactions, like saturated hydrocarbons.
   C. Benzene and compounds containing the benzene-type ring structure are called aromatic compounds. Many have pleasant aromas; others stink.
      1. Today aromatic means compound with rings of electrons and properties similar to benzene. It no longer refers to odor.

9.6   The Chlorinated Hydrocarbons: Many Uses, Some Hazards
   A. Methane derivatives
      1. Methyl chloride (CH_3Cl): used in making silicones.
      2. Methylene chloride (CH_2Cl_2): solvent, paint remover.
      3. Chloroform ($CHCl_3$): early anesthetic; industrial solvent.
      4. Carbon tetrachloride (CCl_4): dry-cleaning solvent.

B. Properties of chlorinated hydrocarbons
    1. Only slightly polar, thus insoluble in water.
    2. These compounds dissolve fats, greases, and oils, thereby making them useful as solvents.
    3. Some chlorinated compounds (DDT, PCBs, etc.) are stored in fatty animal tissue.
C. The Chlorofluorocarbons
    1. CFC's: Carbon compounds containing fluorine, as well as chlorine.
    2. Typical compounds are gases or low-boiling liquids at room temperature.
    3. They are insoluble in water and inert toward most chemical substances.
        a. Used as propellants (for aerosol spray cans).
        b. Their inertness leads to stability in nature, meaning they break down slowly. Diffusion into the stratosphere may damage the ozone layer.
D. Perfluorinated compounds have been used as blood extenders (they dissolve large amounts of oxygen) and in Teflon.
    1. Prefix "per" means all the H atoms have been replaced.

Oxygen-containing Organic Compounds

9.7   See table 9.4, which summarizes the functional groups' names, structure, and general formulas.
    A. Many organic compounds can be divided into two parts:
        1. A functional group.
        2. A hydrocarbon part called an alkyl group.
            a. The letter "R" represents alkyl groups in general structures.
    B. Alkyl groups can be derived from alkanes by removing an H atom from the alkane. They are named by replacing "-ane" with "-yl."
        methyl CH_3—   propyl CH_3—CH_2—CH_2—
        ethyl CH_3—CH_2—   isopropyl CH_3—CH—CH_3—

9.8   The Alcohol Family
    A. An alcohol has a hydroxyl group (OH) substituted for an H of the corresponding alkane.
    B. Methyl alcohol (methanol) is the first member of the family.
        1. Methanol is called wood alcohol and once was made by the destructive distillation of wood.
        2. Methanol, a valuable industrial solvent, is now made from carbon monoxide and hydrogen.
$$CO + 2H_2 \rightarrow CH_3OH$$
        3. Methanol is toxic; it can cause blindness and death.
    C. Ethyl Alcohol (Ethanol)
        1. Ethyl alcohol (ethanol) is also called grain alcohol.
        2. Most ethyl alcohol is made by the fermentation of grain.
    D. Toxicity of Alcohols
        1. Methyl alcohol can cause blindness or death.
        2. Ethyl alcohol is less toxic than methyl alcohol but one pint rapidly ingested can cause death.
        3. Proof is calculated as twice the percent of alcohol.
    E. Multifunctional Alcohols
        1. Some alcohols have more than one hydroxyl (OH) group.

a. Ethylene glycol (antifreeze)

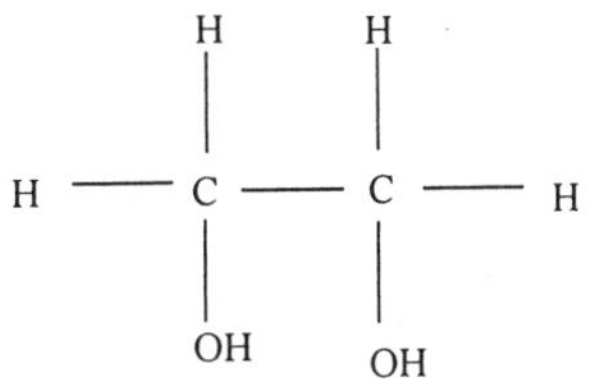

$$
\begin{array}{ccc}
& H & H \\
& | & | \\
H - & C - & C - H \\
& | & | \\
& OH & OH
\end{array}
$$

Ingestion leads to kidney damage and possible death.

b. Glycerol

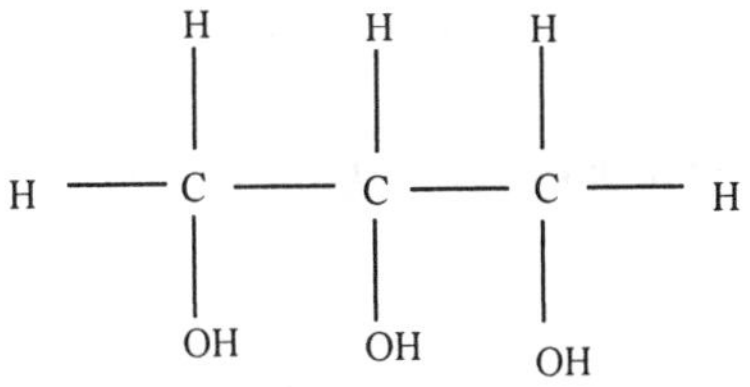

$$
\begin{array}{cccc}
& H & H & H \\
& | & | & | \\
H - & C - & C - & C - H \\
& | & | & | \\
& OH & OH & OH
\end{array}
$$

Used in lotions and nitroglycerin.

9.9 Phenols
    A. Phenols have an —OH group attached directly to a benzene ring.
        1. Phenols are widely used as antiseptics.

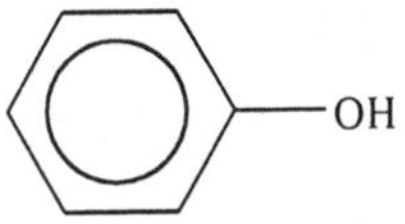

        2. Hydroquinone—benzene ring with hydroxyl (OH) groups at both ends.

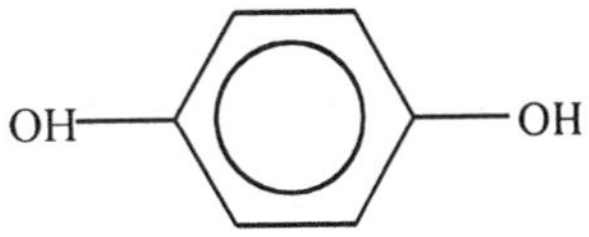

9.10 Ethers
    A. Ethers have two hydrocarbon groups attached to the same oxygen atom (R—O—R′).
        1. Diethyl ether, once used as an anesthetic, is an important solvent. It is highly flammable.
        2. Methyl tert-butyl ether is used as an octane booster in gasoline.

9.11 Aldehydes and Ketones
    A. Both aldehydes and ketones have a carbonyl functional group.

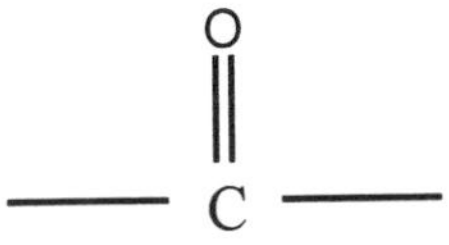

$$
\begin{array}{c}
O \\
\| \\
- C -
\end{array}
$$

B. Aldehydes have a hydrogen atom on the carbonyl carbon atom.

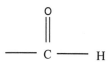

1. The simplest aldehyde is formaldehyde, made by the oxidation of methanol. It is used as a preservative and to make plastics.

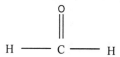

2. The next higher aldehyde is acetaldehyde, made by the oxidation of ethanol.

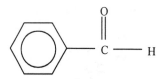

3. Benzaldehyde has an aldehyde group attached to a benzene ring. It is called oil of almond, and is used in flavors and perfumes.

C. Ketones have their carbonyl carbon atom joined to two other carbon atoms. Acetone, a common solvent, is an example.

9.12 Organic Acids
A. Organic or carboxylic acids have the carboxyl group as their functional group.
   1. Carboxyl group

   2. Carboxylic acids are weak acids.
B. Common carboxylic acids include
   1. HCOOH   formic acid   in ant bites, bee stings
   2. CH₃COOH   acetic acid   in vinegar
   3. CH₃CH₂COOH   propionic acid   salts used as preservatives
   4. CH₃CH₂CH₂COOH   butyric acid   rancid butter (stench)
   5. benzoic acid   salts used as preservatives

102

C. Carboxyl acid salts are used as food preservatives (calcium propionate, sodium benzoate).

9.13 Esters: The Sweet Smell of RCOOR′
   A. Esters are derived from carboxylic acids and alcohols.
      1. General formula

$$R - C - OR'$$

   B. Although derived from carboxylic acids with unpleasant odors, many esters have pleasant odors.
      1. They are used in fragrances and flavors.

Nitrogen-Containing Organic Compounds

9.14 Amines and Amides: Nitrogen-Containing Organics
   A. Amines are basic compounds derived from ammonia (NH_3) by replacing one or more of the hydrogen atoms with an alkyl group.
      1. RNH_2, R_2NH, and R_3N are amines.
      2. Amines are named by naming the alkyl groups first, and adding the ending "-amine."
         CH_3NH_2 $CH_3CH_2NH_2$ CH_3NHCH_3
         Methylamine   Ethylamine   Dimethylamine
      3. The simplest aromatic amine has the special name aniline.

      4. The —NH_2 group is called an amino group.
   B. Amides have a nitrogen atom attached to a carbonyl carbon atom.

      1. Nylon, silk, and wool are amides.

9.15 Heterocyclic Compounds: Alkaloids and Others
   A. Heterocyclic compounds are those having atoms other than C in a ring structure.
      1. These compounds usually contain N, O, or S in the ring.
      2. Pyrimidine and purine, found in DNA, are examples.
   B. Alkaloids are amines, often heterocyclic, that occur naturally in plants.
      1. Morphine, nicotine, caffeine, and cocaine are examples.

CHAPTER OBJECTIVES

(You should...)

1. Know that hydrocarbons are compounds of only two elements—hydrogen and carbon.

2. Know three reasons why carbon forms so many compounds.

3. Know that alkanes are hydrocarbons with only single bonds (saturated hydrocarbons).

4. Be able to name the first ten continuous-chain alkanes, given the structure, or to recognize the structure given the name.

5. Be able to convert a structural formula to a condensed structural formula and vice versa.

6. Be able to name cycloalkanes of three to six carbons, given the structure, or to recognize the structure (written out or symbolic, i.e., triangle, square, or pentagon), given the name.

7. Be able to recognize compounds that are homologs (they differ by a CH_2 unit).

8. Know that alkenes (with double bonds) and alkynes (with triple bonds) are unsaturated while alkanes (with single bonds) are saturated.

9. Be able to recognize aromatic hydrocarbons (they contain benzene rings).

10. Be able to name the simplest alkene and alkyne and write their structures.

11. Know that hydrocarbons are used as solvents for fats, oils, waxes, and other substances with low polarity; flammable; insoluble in water; C1–C4 hydrocarbons are gases, C5–C16 hydrocarbons are liquids, the rest are solids; heavier liquid hydrocarbons serve as emollients; and liquid hydrocarbons cause chemical pneumonia.

12. Be able to select compounds that are isomers from a list of structures (they have the same number of each kind of atom).

13. Know that nearly all organic compounds burn (exception: chlorinated compounds such as CCl_4).

14. Be able to recognize a chlorofluorocarbon from its structure and that perfluorocarbons have been used as blood extenders because they readily dissolve oxygen.

15. Be able to recognize the functional group of a(n)

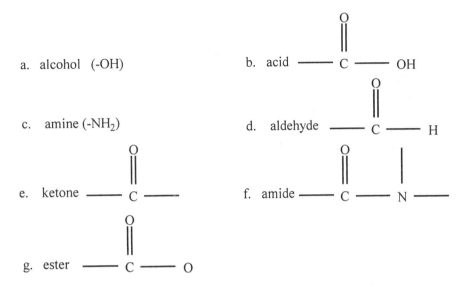

a. alcohol (-OH)

b. acid

c. amine (-NH₂)

d. aldehyde

e. ketone

f. amide

g. ester

16. Be able to name the alcohols, aldehydes, ketones, acids, and amines with one to four carbons (two to four for ketones). Be able to write these structures.

17. Be able to calculate alcohol percent from "proof" and vice versa.

18. Be able to recognize a phenol from its structure and that phenols are used as antiseptics.

19. Be able to name dimethyl ether and diethyl ether, given the structures, or to select the proper structure, given the name.

20. Be able to recognize a heterocyclic compound from its structure.

21. Know that alkaloids are basic compounds (amines) that occur naturally in plants, and be able to name three alkaloids.

22. Be able to recognize the structures for the two multifunctional alcohols, ethylene glycol and glycerol.

DISCUSSION

An understanding of the chemistry of carbon and the hydrocarbons is essential for comprehension of our energy problems and associated pollution problems. Perhaps equally important, the chemistry of the hydrocarbons is the basis for most of the organic and biological chemistry that we will encounter in subsequent chapters. Before going further, you may wish to review covalent bonding in Chapter 5, especially the valence rules. Organic chemistry is very extensive because carbon binds to itself so well and in so many combinations. Focus on the characteristics of the classes (alkanes, alkenes, etc.) of organic compounds to organize your study.

Once again, you may find flash cards helpful in learning the names of the hydrocarbons. Just write the name on one side of a note card and the structure on the other. Look at the name and see if you can write the structure (or vice versa). Then flip the card to see if you got it right.

SELF-TEST

MULTIPLE CHOICE

1. The one element necessarily present in every organic compound is

    a. hydrogen
    c. carbon
    b. oxygen
    d. nitrogen

2. Compounds containing only carbon and hydrogen are known as

    a. methane
    c. carbohydrates
    b. hydrocarbons
    d. aromatic

3. Which of the following is a hydrocarbon?

    a. $C_6H_{13}OH$
    c. $C_6H_{12}O_6$
    b. C_6H_{14}
    d. all are hydrocarbons

4. The proper name for the compound $CH_3CH_2CH_3$ is

    a. ethane
    c. butane
    b. propane
    d. pentane

5. The formula for butane is

    a. CH_3CH_3
    c. $CH_3CH_2CH_2CH_3$
    b. $CH_3CH_2CH_3$
    d. C_6H_6

6. The formula $CH_3CH_2CH_2CH_2CH_3$ represents the compound

    a. propane
    c. hexane
    b. pentane
    d. decane

7. The simplest hydrocarbon compound is

    a. CH_2
    c. CH_3
    b. C_2H_4
    d. CH_4

8. Compounds composed of the same number and kinds of atoms but different in their atomic arrangement are known as

    a. isotopes
    c. homologs
    b. isomers
    d. allotropes

9. A specific series of carbon compounds in which each member differs by CH_2 from the preceding member of the series is known as a(n)

    a. aromatic series   b. homologous series
    c. hydrocarbon series   d. isomeric series

10. In the name cyclohexane, the prefix "cyclo-" means that

    a. the carbon atoms are joined in a ring
    b. the compound is explosive
    c. the compound is a derivative of benzene
    d. each carbon is attached to every other carbon atom

11. Which compound is saturated?

    a. $CH_3 - CH_2 - CH_3$ b. $CH_2 = CH_2$
    c. $HC \equiv CCH_2CH_3$ d. $CHCl = CHCl$

12. Hydrocarbons such as $CH_2=CH_2$ and $CH_3CH=CH_2$ are said to be
    a. aromatic   b. saturated
    c. unsaturated   d. ionic

13. The proper name for the compound is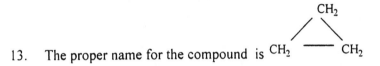

    a. propane   b. pentane
    c. cyclopropane   d. cyclopentane

14. Cyclopentane is correctly represented by

    a. CH₃CH₂CH₃   b. CH₃CH₂CH₂CH₂CH₃

    c.   d.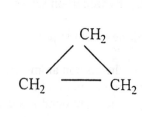

15. Which of the following is an aromatic hydrocarbon?

    a. CH₃ —— CH₂ —— CH₃
    b. CH₂ ═══ CH₂
    c. HC ≡≡≡ CCH₂CH₃

    d.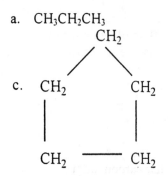

16. The correct structure for ethylene (C_2H_4) is

a.
```
H - C - C - H
    |   |
    H   H
```

b.
```
H - C = C - H
    |   |
    H   H
```

c.
```
      H
      |
H - C - C - H
      |
      H
```

d.
```
H - C ≡ C - H
    |       |
    H       H
```

17. Choose the statement that best describes hydrocarbons.

   a. Flammable ionic compounds that are insoluble in water and are more dense than water.
   b. Flammable ionic compounds that are soluble in water and are more dense than water.
   c. Nonflammable covalent compounds that are insoluble in water and are less dense than water.
   d. Flammable covalent compounds that are insoluble in water and are less dense than water.

18. Acetylene is a(n)

   a. alkane   b. alkene
   c. alkyne   d. aromatic compound

19. Which is not a gas?

   a. methane   b. ethene
   c. acetylene   d. octane

20. If hexane and water are mixed, the result is

    a. a clear solution of hexane dissolved in water
    b. a layer of hexane sitting on top of a layer of water
    c. a layer of water sitting on top of a layer of hexane
    d. an explosion

21. Benzene is often represented as

    a. $CH_3CH_2CH_2CH_3$

    b. $HC \equiv CCH_2CH_2C \equiv CH$

    c.

    d.

22. Which of the following are isomers?

I

$CH_3CH_2CH_2CH_2CH_3$

II

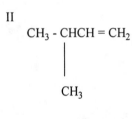

III

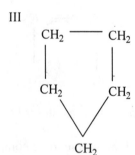

    a. all three   b. I and II
    c. I and III   d. II and III

23. The compounds $CH_3CH_2CH_3$ and $CH_3CH_2CH_2CH_3$ are

    a. allotropes   b. homologs
    c. isomers   d. isotopes

24. Which of the following would not burn?

    a. CH_4 b. CH_3OH
    c. CCl_4 d. $CH_3CH_2OCH_2CH_3$

25. Which of the following is a chlorofluorocarbon?

    a. $CHCl_3$ b. $CFCl_3$
    c. C_3F_8 d. CHF_3

26. Which of the following is chloroform?

    a. CH_3Cl b. CH_2Cl_2
    c. $CHCl_3$ d. $CFCl_3$

27. The compound CH_3CH_2Cl is named

    a. carbon dichloride   b. methyl chloride
    c. methylene chloride   d. ethyl chloride

28. Which alcohol is the least dangerous to drink?

    a. methyl alcohol   b. ethyl alcohol
    c. isopropyl alcohol   d. all are equally dangerous

29. Isopropyl alcohol is

    a. $\underset{\underset{\displaystyle OH}{|}}{CH_3CHCH_3}$ b. CH_3OH

    c. CH_3CH_2OH d. $CH_3CH_2CH_2OH$

30. Grain alcohol in the United States is made

    a. from coal   b. by fermenting starches and sugars
    c. from vinegar   d. by fermenting acetic acid

31. A combination of atoms that confers certain chemical and physical properties on a compound is called a (n)

    a. ether   b. functional group
    c. hydrogen bond   d. homolog

32. Which is an alcohol?

    a. C_6H_6 b. $\underset{\underset{\displaystyle O}{\|}}{CH_3\text{-}C\text{-}CH_3}$

    c. $CH_3\!\!-\!\!O\!\!-\!\!CH_3$ (d.) $CH_3CH_2CH_2OH$

33. Which compound is an ether?

    a. $\overset{\overset{\displaystyle O}{\|}}{CH_3C}\text{-}OCH_3$ b. CH_3CH_2OH

    (c.) $CH_3OCH_2CH_3$ d. $\overset{\overset{\displaystyle O}{\|}}{CH_3C}\text{-}CH_3$

34. The correct name for CH₃—O—CH₃ is

    a. dimethyl ether   b. ethyl ether
    c. diethyl ether   d. oxyethane

35. Vodka that is 90 proof contains what percent alcohol?

    a. 9   b. 45
    c. 90   d. 180

36. Wood alcohol is the same as

    a. methanol   b. 2-propanol
    c. grain alcohol   d. rubbing alcohol

37. Phenol is

    a. OH   b. OH

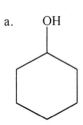 

    c. CH₃CH₂CH₂OH   d. CH₃CH₂CH₂CH₂CH₂OH

38. The formula for diethyl ether is

    a. CH₃OCH₃   b. C₂H₅OH

                   O
                   ‖
    c. CH₃CH₂CCH₂CH₃   d. CH₃CH₂OCH₂CH₃

39. The alcohol present in alcoholic beverages is

    a. methyl alcohol   b. ethyl alcohol
    c. denatured alcohol   d. wood alcohol

40. Denatured alcohol refers to

    a. any alcohol not produced by fermentation
    b. grain alcohol that is highly taxed
    c. ethyl alcohol that has been treated with something to make it unfit to drink
    d. methanol

41. The ingestion of methyl alcohol can result in

    a. deafness
    b. blindness
    c. severe depression
    d. all of these

42. When ingested, ethanol acts as a(n)

    a. stimulant
    b. depressant
    c. antiseptic
    d. analgesic

43. A group that aldehydes and ketones have in common is

    a. —COOH
    b. $-C=O$
       |
       H

    c. $-C=O$
       |
    d. —OH

44. The name of the functional group of aldehydes and ketones is the

    a. carbonyl group
    b. carboxyl group
    c. double bond
    d. hydroxyl group

45. The representation –COOH is called a

    a. carboxyl group
    b. carbonyl group
    c. aldehyde group
    d. hydroxyl group

46. Which acid is found in vinegar?

    a. formic acid
    b. nitric acid
    c. propionic acid
    d. acetic acid

47. The product of the reaction between an alcohol and a carboxylic acid is known as a(n)

    a. acid anhydride
    b. ester
    c. ether
    d. salt

48. Alkaloids can best be generally classified with the

    a. alcohols
    b. acids
    c. amines
    d. amides

MATCHING

Match each structure with an acceptable name. Caution: There are more names than structures.

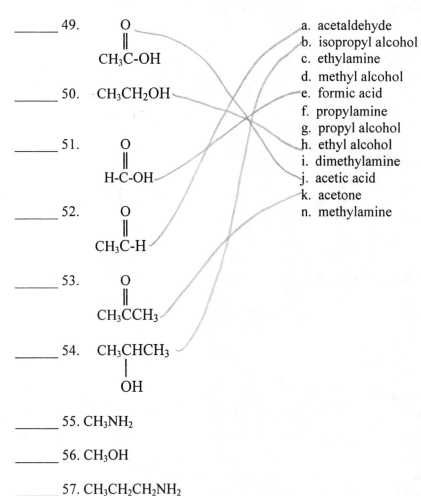

_____ 49.

$$CH_3\overset{\displaystyle O}{\overset{\|}{C}}-OH$$

_____ 50. CH_3CH_2OH

_____ 51.

$$H-\overset{\displaystyle O}{\overset{\|}{C}}-OH$$

_____ 52.

$$CH_3\overset{\displaystyle O}{\overset{\|}{C}}-H$$

_____ 53.

$$CH_3\overset{\displaystyle O}{\overset{\|}{C}}CH_3$$

_____ 54. CH_3CHCH_3
$\quad\quad\quad\quad |$
$\quad\quad\quad\quad OH$

_____ 55. CH_3NH_2

_____ 56. CH_3OH

_____ 57. $CH_3CH_2CH_2NH_2$

_____ 58. CH_3NHCH_3

a. acetaldehyde
b. isopropyl alcohol
c. ethylamine
d. methyl alcohol
e. formic acid
f. propylamine
g. propyl alcohol
h. ethyl alcohol
i. dimethylamine
j. acetic acid
k. acetone
n. methylamine

ANSWERS

1. c	10. a	19. d	27. d	35. b	43. c	51. e
2. b	11. a	20. b	28. b	36. a	44. a	52. a
3. b	12. c	21. c	29. a	37. b	45. a	53. k
4. b	13. c	22. d	30. b	38. d	46. d	54. b
5. c	14. c	23. b	31. b	39. b	47. b	55. n
6. b	15. d	24. c	32. d	40. c	48. c	56. d
7. d	16. b	25. b	33. c	41. b	49. j	57. f
8. b	17. d	26. c	34. a	42. b	50. h	58. i
9. b	18. c					

Polymers

Giants among Molecules

KEY TERMS

addition polymers	monomer	silicone
celluloid	plasticizers	thermoplastic polymers
condensation polymerization	polyamide	thermosetting polymers
copolymer	polyester	vulcanization
elastomer	polymer	

CHAPTER SUMMARY

10.1 Polymerization: Making Big Ones Out of Little Ones
    A. Polymer comes from the Greek term for "many parts."
        1. Polymers are macromolecules.
        2. The building blocks of polymers are called monomers ("one part").
    B. Polymerization is the process by which monomers are converted to polymers.

10.2 Natural Polymers
    A. Starch is a polymer made up of glucose units.
        1. Cotton and wool are also polymers of glucose.
    B. Proteins are polymers of amino acids.
    C. Nucleic acids are polymers in DNA.

10.3 Celluloid: Billiard Balls and Collars
    A. Celluloid is derived from natural cellulose by reacting it with nitric acid to form cellulose nitrate.
    B. Celluloid is seldom used today because it is highly flammable.

10.4 Polyethylene: From the Battle of Britain to Bread Bags
    A. Polyethylene is made from ethylene (CH_2—CH_2), a hydrocarbon derived from petroleum.
    B. There are three major types of polyethylene plastics.
        1. High-density polyethylene (HDPE) is made from linear polymer strands that have great rigidity and strength. These plastics have greater rigidity and tensile strength.
            a. Used in toys and gallon jugs.

2. Low-density polyethylene (LDPE) is made from branched chains that cannot pack as tightly, thus producing a more flexible material.
   a. Used in squeeze bottles and insulation.
3. Linear low-density polyethylene (LLPE) is a copolymer of ethylene and a higher alkane like 1-hexene.
   a. Used to make plastic film.
C. Polyethylene is a <u>thermoplastic</u> material; it can be softened by heat and then reformed.
   1. Some plastics are <u>thermosetting plastics,</u> which means that they harden permanently when formed.

10.5 Addition Polymerization: One + One + One ... Gives One!
   A. There are two general types of polymerization reactions.
      1. Addition polymerization uses an addition reaction in such a way that the polymeric product contains all of the atoms of the starting monomers.
      2. Condensation polymerization results in a product where part of the monomer molecule is not included.
   B. The repeating units in a polymer structure are called segmers.
   C. Examples of addition reactions
      1. Polypropylene—polymer of propylene with methyl groups on every other carbon.
         a. Used in hard-sided luggage
      2. Polystyrene—polymer of styrene (benzene rings with ethylene attached).
         a. Used in disposable cups.
      3. Vinyl Polymers—polymer of vinyl chloride (ethylene molecule with a chlorine attached) resulting in polyvinyl chloride (PVC).
         a. Used in plastic wrap and clear plastic bottles.
      4. Teflon: the nonstick coating
         a. Polymer of tetrafluorethylene (ethylene with all H's substituted with F's).
            i. Used in electrical insulation and gaskets.
   D. Polymerization Equations
      1. Repeating unit called a segmer is written inside brackets [ ] with bonds extending from brackets in both directions.
         a. Subscript of "n" indicates the segmer is repeated an unspecified number of times.
   E. Polyacetylene
      1. Acetylene molecules have triple bonds and can undergo addition reactions.
         a. Result is a polymer with every other bond a double bond.
         b. The alternating single and double bonds in this polymer make it possible to conduct electricity.

10.6 Rubber
   A. The plastics industry grew out of the need for a substitute for natural rubber during World War II.
   B. Natural rubber is a polymer of isoprene.

   C. Vulcanization is a process cross-linking the hydrocarbon chains of natural rubber with sulfur atoms.
      1. Vulcanized rubber is harder.
      2. Elastomers are stretchable materials that return to their original structure.
   D. Synthetic Rubber
      1. Some synthetic elastomers are polybutadiene, polychloroprene (Neoprene), and styrene-butadiene rubber (SBR).

2. Copolymers incorporate two different monomer units into a polymer chain. SBR is an example.
   a. Copolymers are more resistant to oxidation and natural abrasion but have less satisfactory mechanical properties.
E. Polymers are added to paint to help harden it into a continuous surface coating.
   1. This polymer is called a binder.

10.7 Condensation Polymers: Splitting Out Water
A. In condensation polymers, a portion of the monomer molecule is not incorporated into the final polymer.
B. Nylon-6 is formed by splitting out water to join 6-aminohexanoic acid molecules into long polyamide chains.
   1. An amide bond is formed between the carbonyl of one monomer and the amide group of another.
   2. Water molecules are the by-products.
   3. Nylon is a polyamide—amide linkages hold the molecule together.
C. Polyesters (Dacron)

$$\left[\begin{array}{c} O\text{-}CH_2\text{-}CH_2\text{-}O\text{-}\overset{\displaystyle O}{\overset{\|}{C}}\text{---}\bigcirc\text{---}\overset{\displaystyle O}{\overset{\|}{C}} \end{array}\right]$$

   1. Hydroxyl groups of ethylene glycol react with carboxylic groups in terephthalic acid to form ester linkages.
D. Bakelite is a stable, thermosetting, condensation polymer that won't soften on heating.
   1. It is made from phenol and formaldehyde.
   2. It is a three-dimensional polymer composed of phenol units, each with three formaldehyde cross-links.
   3. Thermosetting resins become permanently hard; they cannot be softened and remolded.
E. Other Condensation Polymers
   1. Polycarbonates—"tough as glass" polymers.
      a. Used in bullet-proof windows.
   2. Polyurethane
      a. Used in foam rubber.
   3. Epoxy resins
      a. Used in sports gear and car bodies.
F. Composites
   1. High strength fibers (glass, graphite or ceramics) are held together with a polymeric matrix (usually a thermoset condensation polymer).
   2. Composites have the strength of steel but weigh only a fraction of the weight.
      a. Powerful adhesives form when two components are mixed and polymer chains become cross-linked.
G. Silicones—polymers of alternating silicon and oxygen atoms
   1. Very heat stable and resistant to most chemicals.
   2. Used as water-proofing materials, "Silly Putty," and synthetic body parts.

10.8 Properties of Polymers
A. Crystalline and Amorphous Polymers
   1. Crystalline ploymers: molecules line up to form long fibers of high tensile strength.

2. Amorphous polymers: randomly orientated molecules that tangle with one another providing good elasticity.
3. Two molecular structures can be grafted onto one polymer chain resulting in both sets of properties — flexibility and rigidity.
B. Glass transition temperature (T_g)
   1. Polymers are rubbery above their glass transition temperature (T_g) and are hard and brittle, like glass, below their T_g.
C. Fiber-forming properties
   1. Three-fourths of fibers and fabrics used in the United States are synthetic.
      a. Synthetic fibers can mimic silk, wool, or enhance the properties of cotton by making it wrinkle-resistant.
      b. Some fibers (Dyneema) rival steel wire in strength.

10.9   Disposal of Plastics
A. Most plastics break down slowly in the environment, leading to litter and solid waste problems.
   1. When disposed of in oceans, plastics pose a danger to fish by clogging their digestive tracts.
B. Many plastics can be recycled, but they must be separated according to type.
   1. Code numbers specifying the type of plastic and stamped on the bottom of plastic containers help in the separation process.
      a. Only about 2% of waste plastic is currently recycled.
C. Incineration of plastics could supply a good source of energy.
   1. Incineration can lead to air pollution and clogging of incinerators by melted plastics.
D. Degradable Plastics
   1. Most are starch-based synthetic polymers with light-sensitive additives.

10.10   Plastics and Fire Hazards
A. Some synthetic fabrics are especially flammable.
B. The fire retardant "Tris" has been shown to be carcinogenic and mutagenic.
   1. Mutagens are substances that cause changes in DNA. Mutations are passed on to succeeding generations.
C. Plastics form toxic gases when burned, presenting special hazards to fire fighters and occupants of burning buildings.
   1. Hydrogen cyanide (HCN) is released by burning polyacrylonitriles such as Orlon.

10.11   Plasticizers and Pollution
A. Hard and brittle polymers can be made more flexible by the addition of chemicals called plasticizers.
   1. Plasticizers lower the T_g of the plastic.
B. Plasticizers are usually lost by evaporation as the plastic ages, leaving the plastic brittle.
C. Polychlorinated biphenyls (PCBs) have been banned from use as plasticizers because they degrade slowly in nature.
   1. PCB's resemble DDT in structure.
   2. PCB residues have been found in animals and water supplies.
   3. PCB's have high electrical resistance, which makes them useful in electrical apparatus.
D. Phthalate esters are the most widely used plasticizers today.

10.12   Plastics and the Future
A. Synthetic polymers conduct electricity, perform as adhesives, and can be stronger than steel but lighter in weight.

B. Plastics have served as artificial body part materials.
C. Plastics are made from non-renewable petroleum

CHAPTER OBJECTIVES

(You should ...)

1. Be able to match the monomers ethylene —[$CH_2 = CH_2$]—; vinyl chloride—[$CH_2 = CHCl$]—; tetrafluoroethylene—[$CF_2 = CF_2$]—; acrylonitrile—[$CH_2 = CHCN$]—; and styrene—[$CH = CH_2$]—with the polymer that each forms.

2. Be able to match the polymer names polyethylene, polyvinyl chloride, polyacrylonitrile (Orlon), polytetrafluoroethylene (Teflon), and polystyrene with the appropriate structure.

3. Know that polymers are composed of giant molecules (compared to water, methane, etc.), yet these molecules are still usually microscopically small.

4. Know the difference between addition and condensation polymers.

5. Know that polyethylene is the most common synthetic polymer. It is used to make bread bags, fruit bags, garbage can liners, etc.

6. Know that Teflon is used to coat nonstick cookware.

7. Know that vinyl objects are made from polyvinyl chloride.

8. Know that polystyrene is used to make Styrofoam objects.

9. Know that ethylene molecules have carbon-to-carbon double bonds but that polyethylene does not.

10. Be able to recognize a polyamide by its structure.

11. Know that polyesters are condensation polymers.

12. Know that nylon is a polyamide. (Proteins are also polyamides.)

13. Know that Dacron is a polyester.

14. Know that sulfur is added to rubber to make it harder (by cross-linking the polymer chains.)

15. Know that cross-linking of polymers results in a more rigid material.

16. Know that Bakelite is a cross-linked polymer.

17. Be able to give the structure of a polymer molecule four segmers long, provided you are given the structure of the monomer(s).

18. Know that 75% of the fibers used in the United States today are synthetic.

19. Know that a polymer is rubbery and tough at temperatures above its glass transition temperature (T_g) and hard and brittle below its T_g.

20. Know that plasticizers lower the T_g of a polymer.

21. Know that plasticizers are substances that make otherwise brittle polymers flexible.

22. Know that polychlorinated biphenyls and phthalate esters have been or are currently used as plasticizers.

23. Know that disposal of plastics is complicated by the fact that most are not usually biodegradable.

24. Know that thermoplastics can be melted down and remolded.

25. Know that Orlon forms hydrogen cyanide (HCN) when burned.

DISCUSSION

This can be called the Plastic Age (as contrasted to the Stone Age or Iron Age). Humanity has used natural polymers (rubber, wood, cotton) for millennia. Since the early 1900's, chemists have been able to produce polymers and tailor their properties by modifying their structure. Since polymers ("many units") are made out of monomers ("one unit"), the structure of the polymer is determined by the structure of the monomer.

We classify polymers by several methods:
1. Thermoplastic—can be reformed by softening with heat.
   Thermosetting—once made, these cannot be reformed.
2. Addition polymers—monomer needs a double (or triple) bond .
   Condensation polymers—monomer needs two functional groups and a small molecule is "condensed" out.
2. Low-density polymers—loosely packed polymer strands; soft and flexible.
   High-density polymers—tightly packed polymer strands; harder and rigid.

You should concentrate on relating the monomer structure to the polymer it makes and the properties of the polymer. The properties of the polymers are determined by their monomers. These properties can be modified by introducing additives such as plasticizers, flame retardants, and fibers (to make composites). While there are many advantages to polymers, fire hazards and disposal issues are a problem. The research in polymers is very active with new developments occurring frequently.

SELF-TEST

Multiple Choice

1. The small-molecule starting materials from which polymers are constructed are called

   a. hydrocarbons
   c. segmers
   b. monomers
   d. plasticizers

2. A thermoset plastic is a substance that

    a. will return to its original shape after being stretched
    b. softens on heating and can be molded under pressure
    c. hardens under the influence of heat and pressure
    d. is a cheap imitation of natural materials

3. Which compound would not serve as a monomer for addition polymerization?

    a. $CH_2=CHCN$ b. $CH_2=CHCl$
    c. $CH_2=CHCOOH$ d. $HOCH_2COOH$

4. Styrofoam picnic coolers are made from

    a. $CH_2=CH_2$ b. $CH_2=CHCl$
    c. $CH_2=CHCN$ d. $CH_2=CH—$

5. The nonstick coating on cookware is made from

    a. $CH_2=CH_2$ b. $CH_2=CHCl$
    c. $CH_2=CHCN$ d. $CF_2=CF_2$

6. A high-density polyethylene is most likely to be used in

    a. bread bags   b. toys
    c. plastic film   d. squeeze bottles

7. Vinyl hardtops for automobiles are made from

    a. $CH_2=CH_2$ b. $CH_2=CHCl$
    c. $CH_2=CHCN$ d. $CF_2=CF_2$

8. Plastic bags are usually made from

    a. $CH_2 = CH_2$

    b. $CH_2 = CHCl$

    c. $CH_2 = CHCN$

    d. $CH_2 = CH$

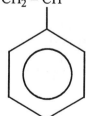

9. Natural rubber is

   a. polyethylene
   b. Neoprene
   c. polyisoprene
   d. polyvinyl chloride

10. Which, if any, are <u>not</u> true of the vulcanization process?

   a. It results in sulfur bridges between polymer chains
   b. It increases the hardness of natural rubber
   c. It improves the elasticity of natural rubber
   d. All of the above are true of the vulcanization process

11. The polymer formed from $CH_2 = C - CH_3$ with Cl attached

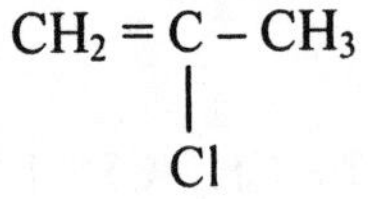

   a.
   $$\left[CH_2 = \underset{\underset{Cl}{|}}{\overset{\overset{CH_2Cl}{|}}{C}} \right]_n$$

   b.
   $$\left[CH_2 = \underset{\underset{Cl}{|}}{\overset{\overset{CH_2Cl}{|}}{C}} - CH_2 \right]_n$$

   c.
   $$\left[CH_2 - \underset{\underset{CH_3}{|}}{\overset{\overset{Cl}{|}}{C}} \right]_n$$

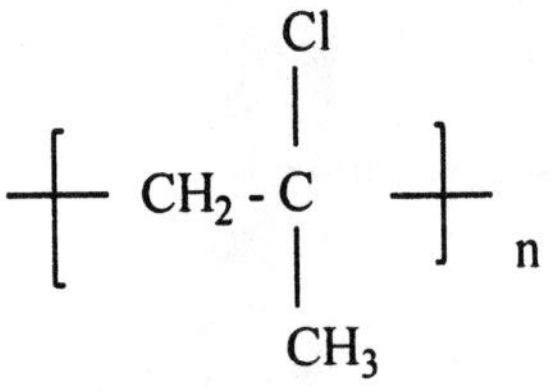

   d.
   $$\left[CH - \underset{\underset{Cl}{|}}{\overset{\overset{CH_3}{|}}{CH}} \right]_n$$

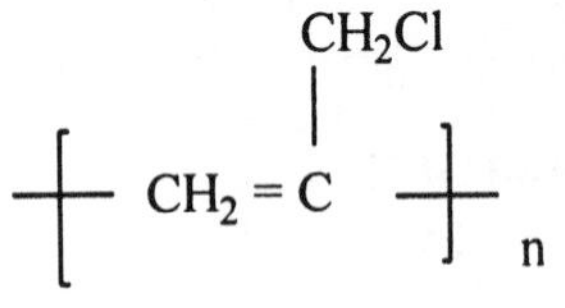

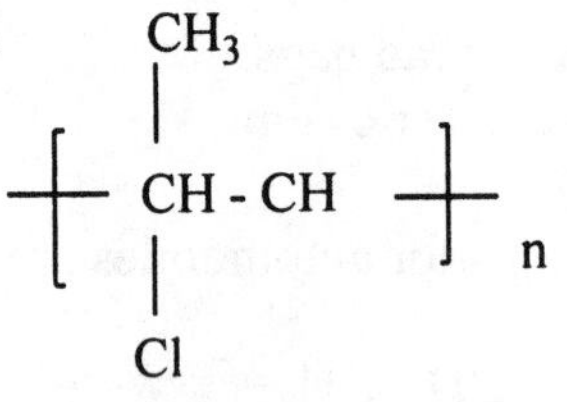

12. If the monomer is $CH_3 - C = CH - Cl$, the polymer is

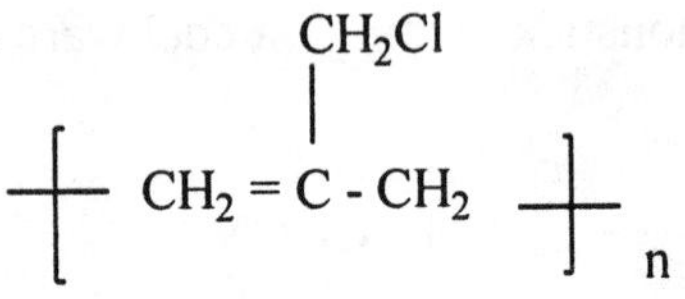

   a.
   $$\left[CH_3 - CH = \underset{\overset{Cl}{|}}{CH} \right]_n$$

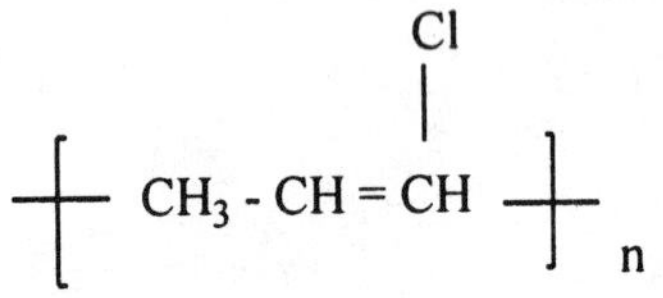

   b.
   $$\left[CH_3 - CH_3 - \underset{\overset{Cl}{|}}{CH} \right]_n$$

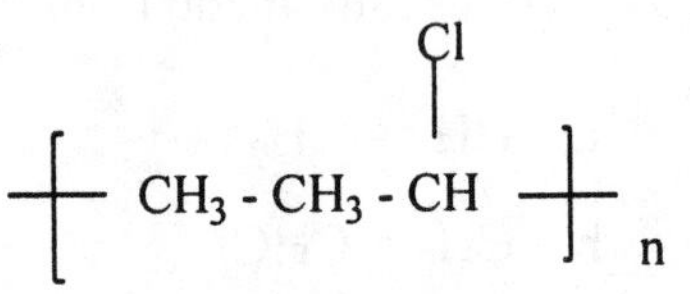

   c.
   $$\left[CH_2 - CH = \underset{\overset{Cl}{|}}{CH} \right]_n$$

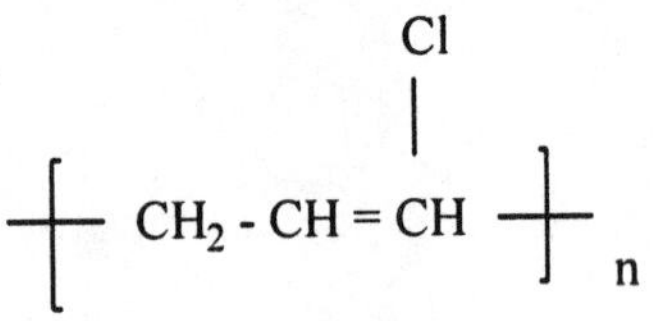

   d.
   $$\left[CH - \underset{\underset{Cl}{|}}{\overset{\overset{CH_3}{|}}{CH}} \right]_n$$

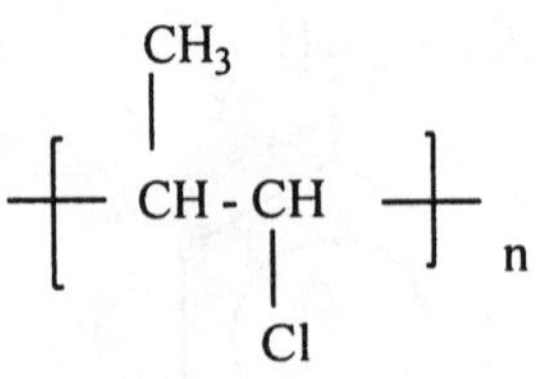

13. Adding halogens to an addition monomer

   a. increases its resistance to flame
   b. changes the color to brown
   c. makes the polymer smell
   d. all of the above

14. Which compound should be a monomer for a condensation polymer?

    a. $H_2C=CH_2$ b. $HOOC—CH_3$
    c. $CH_3—NH_2$ d. $HOOCNH_2$

15. Which does <u>not</u> apply to the polymer Bakelite?

    a. It is a copolymer   b. It is a condensation polymer
    c. It is a cross-linked polymer   d. It is an elastomer

16. Dacron is a

    a. polyamide   b. polyester
    c. polyolefin   d. acrylic fiber

17. Nylons are

    a. polyamides   b. polyesters
    c. polyolefins   d. acrylic fibers

18. What is the functional group linking the segmers of nylon?
    a. amide   b. ester
    c. $C=C$ d. $C=O$

19. Plasticizers are

    a. polymers that have elastic properties
    b. polymers that soften on heating
    c. molecules that make brittle polymers flexible
    d. extremely toxic

20. Which compounds are currently used as plasticizers?

    a. phthalate esters   b. polychlorinated biphenyls
    c. vinyl chlorides   d. polypropylene

21. Disposal of synthetic plastics is complicated by

    a. the large volume of these materials that accumulates as trash
    b. the non-biodegradable nature of most synthetic plastics
    c. the toxicity of some gases produced when the materials are burned
    d. all of the above are valid

22. What percentage of fibers used in the United States today are synthetic?

    a. 5   b. 25
    c. 75   d. 90

23. The temperature at which a polymer changes from hard and brittle to rubbery and tough is called its

    a. boiling point   b. melting point
    c. glass transition temperature   d. thermosetting temperature

24. Burning polyacrylonitrile (Orlon) generates
    a. O_2 b. HCl
    c. HCN   d. N_2O

25. PCB's have been banned as plasticizers because
    a. they are acutely toxic
    b. their residues have been found in animals, water, and sediments
    c. they are mutagens
    d. they degrade too quickly to be useful

26. Composites are formed by adding_____ to condensation polymers

    a. flame retardants   b. addition polymers
    c. fibers   d. plasticizers

27. Silicones are formed from

    a. carbon-silicon compounds   b. sand
    c. silicon-oxygen-carbon compounds   d. silicone-halogen compounds

28. Advantages of silicones are

    a. heat stable   b. chemical resistant
    c. water-proof   d. all of the above

29. The hardest step in recycling plastics is

    a. collecting   b. chopping
    c. melting   d. remolding

30. Half of the waste from polymers is

    a. tires   b. packaging material
    c. credit cards   d. VCR tapes

31. The greatest hazard of polymers in fires is

    a. the heat produced   b. the oxygen consumed
    c. the toxic gases released   d. all are equally important

MATCHING

Match the polymer with the monomer (small letter) from which it is made and with the appropriate name (capital letter).

_____ 32. • • • CH$_2$ - CHCH$_2$CHCH$_2$ - CH • • •   a. CH$_2$ = CH   A. polyethylene

$\quad\quad\quad\quad\quad\quad\quad\quad$ | $\quad$ | $\quad$ |

$\quad\quad\quad\quad\quad\quad\quad\quad$ Cl $\quad$ Cl $\quad$ Cl

_____ 33. • • • CH$_2$CHCH$_2$CHCH$_2$ CH • • •   b. CF$_2$ = CF$_2$ B. polyacrylonitrile (Orlon)

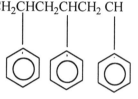

_____ 34. • • • CH$_2$CHCH$_2$CHCH$_2$CH • • •   c. CH$_2$ = CHCl   C. polytetrafluoroethylene (Teflon)

$\quad\quad\quad\quad\quad\quad$ | $\quad$ | $\quad$ |

$\quad\quad\quad\quad\quad\quad$ CN $\quad$ CN $\quad$ CN

_____ 35. • • • CH$_2$CH$_2$CH$_2$CH$_2$CH$_2$CH • • •   d. CH$_2$ = CHCN   D. polyvinyl chloride

_____ 36. • • • CF$_2$CF$_2$CF$_2$CF$_2$CF$_2$CF$_2$ • • •   e. CH$_2$ = CH$_2$ E. polystyrene

ANSWERS

1. b	7. b	13. a	19. c	25. b	31. c
2. c	8. a	14. d	20. a	26. c	32. c, D
3. d	9. c	15. d	21. d	27. c	33. a, E
4. d	10. d	16. b	22. c	28. d	34. d, B
5. d	11. c	17. a	23. c	29. a	35. e, A
6. b	12. d	18. a	24. c	30. b	36. b, C

Chemistry of the Earth

Metals and Minerals

KEY TERMS

alloy	ceramic	slag
asbestos	glass	steel
bronze	micas	synergistic effect
cement	quartz	

CHAPTER SUMMARY

11.1   Spaceship Earth: The Materials Manifest
   A. Structurally, the Earth is divided into three main regions.
      1. The core is thought to be mainly iron and nickel.
      2. The mantle is believed to consist mainly of silicates (sulfur, oxygen, and metal compounds).
      3. The crust is the outer shell of the Earth; it has three parts.
         a. The lithosphere is the solid part.
         b. The hydrosphere is the oceans, lakes, rivers, etc.
         c. The atmosphere (air) is the gaseous part.
   B. The most abundant element in the Earth's crust is oxygen, followed by silicon.

11.2   The Lithosphere: Organic and Inorganic
   A. Prominent minerals in the lithosphere
      1. Silicates: metals combined with silicon and oxygen.
      2. Carbonates: metals combined with carbon and oxygen.
      3. Sulfides: metals combined with sulfur.
   B. Minerals make up the inorganic portion of the earth's crust.
   C. The organic portion of the lithosphere includes all living things, their waste and decomposition products, and fossilized materials.

11.3   Meeting Our Needs: From Sticks to Bricks
   A. Clay is converted to ceramic materials (bricks, pottery) by heating to a high temperature.
   B. Fire was one of the earliest agents of chemical change discovered by man.
   C. With high temperatures, people were able to make glass and separate metals from their ores.

11.4   Silicates and the Shape of Things
   A. The basic silicate unit is the SiO_4 tetrahedron.

1. Silicates can be arranged linearly in fibers, in planar sheets, or in complex three-dimensional arrays.

B. In quartz, each sulfur atom is surrounded by four oxygen atoms.
   1. Amethyst, citrine, rose, and smoky quartz contain impurities, which give them their characteristic color.

C. Micas are composed of sheets of tetrahedra, giving a two-dimensional structure.

D. Asbestos is a generic term for fibrous silicates.
   1. Chrysotile is a form of asbestos composed of a double chain of SiO_4 tetrahedra bonded to magnesium ions.
   2. Inhalation of asbestos fibers over 10 to 20 years causes asbestosis, a severe respiratory disease. After 30 to 45 years, cancer may develop.
   3. Cigarette smoking and inhalation of asbestos fibers act synergistically to greatly increase the risk of lung cancer.
   4. Harmful effects of asbestos are due to crocidolite, a rare form of asbestos.

11.5   Modified Silicates: Ceramics, Glass, and Cement
   A. Ceramics (modified clays) have been developed with specialized properties such as high heat resistance, magnetic properties, and computer memory capabilities.
      1. Superconductors are ceramics that have exceptional electrical conductivity at very low temperatures.
   B. Clay is basically aluminum silicate.
   C. Glass is made by heating a mixture of sand, sodium carbonate, and limestone (calcium carbonate).
      1. Glass is characterized by an irregular three-dimensional array of SiO_4 tetrahedra.
         a. Glass softens due to heat breaking the weaker bonds.
   D. Cement and concrete are made by heating a mixture of limestone and clay.

11.6   Metals and Ores
   A. Copper and Bronze
      1. Copper was probably the first metal to be separated from ore through smelting.
      2. Bronze, a copper-tin alloy, is harder than copper.
   B. Iron and Steel
      1. Carbon reduces iron oxides to iron metal. (The actual reducing agent is carbon monoxide.)
      2. Iron is converted to steel by reacting oxygen with the impurities in iron and adjusting the carbon content.
         a. High-carbon steel is hard and strong.
         b. Low-carbon steel is ductile and malleable.
   C. Aluminum: Abundant and Light
   D. Environmental cost of iron and aluminum
      1. Aluminum is the most abundant metal in the Earth's crust, occurring as aluminum oxide (bauxite).
      2. Making an aluminum can from bauxite requires 6.3 times as much energy as making a steel can.
      3. When aluminum is recycled, only a fraction of the energy needed to make a steel can is required.
   E. Other important metals
      1. Table 11.3 (in your text) lists other important metals such as chromium, gold, and magnesium.

2. Although atoms are conserved globally, use of metals scatters these atoms throughout the environment.

11.7 Running Out of Everything: Earth's Dwindling Resources
   A. Most of Earth's high-grade ores are gone.
   B. It takes more energy to obtain metals from low-grade ores than from high-grade ores.

11.8 Land Pollution: Solid Wastes
   A. Open dumps become infested with rodent and insect pests; they are being phased out in the United States.
   B. In sanitary landfills, trash is compacted and covered over. Land for landfills is increasingly scarce.
   C. Incineration can reduce the volume of trash, but it can lead to air pollution.
   D. New methods of solid waste disposal include making useful materials from refuse.

11.9 The Three R's of Garbage: Reduce, Reuse, Recycle
   A. Average person discards 1.5 kg of trash each day.
   B. Reduce amount of throwaway material.
   C. Reuse materials: Durable objects can be used repeatedly and require less energy than recycling.
   D. Recycle materials: This requires less energy than making objects from ores.

11.10 How Crowded Is Our Spaceship?
   A. Due to medical advances, our death rate has been lowered while our birthrate has stayed the same.
      1. As a result, our birth and death rates are out of balance and our population is increasing at a very fast rate.
         a. This creates a strain on our resources.

CHAPTER OBJECTIVES

(You should know that...)

1. The mass of the Earth is essentially constant; material substances are not being replenished.

2. Oxygen is the most abundant element on the surface of the Earth.

3. A metal is usually obtained from its ore by reduction of the ore.

4. The three parts of the Earth's crust are the lithosphere, hydrosphere, and atmosphere.

5. Mica, which is found as thin sheets, has SiO_4 tetrahedra arranged in a flat, two-dimensional array.

6. Asbestos, a fibrous mineral, has chains of SiO_4 tetrahedra.

7. The three principal raw materials for making glass are sand, sodium carbonate, and limestone.

8. The two principal raw materials for making cement are limestone and clay.

9. Asbestos and cigarette smoke act synergistically to cause cancer.

10. The properties of steel may be adjusted by varying the carbon content.

11. Clay, quartz, asbestos, mica, glass, ceramics, and cement are silicates or modified silicates.

12. The special properties of glass, such as softening when heated rather than melting sharply, are due to the irregular arrangement of SiO_4 tetrahedra.

13. The lower the grade of an ore, the more energy is required to concentrate it.

14. Use of throw-away bottles requires 10 times as much energy as the use of returnables, which average 12.5 uses before loss or breakage.

15. Production of an aluminum can requires 6.3 times as much energy as production of a steel ("tin") can.

16. The basic raw materials for the production of iron are iron ore, coal, and limestone.

17. Although atoms are conserved, a substance may be effectively lost through scattering in the environment.

18. The United States is highly dependent on foreign countries for several strategic metals and minerals.

19. Mining in the United States is primarily for low-grade ores. Such mining requires a lot of energy.

20. Open dumps present the threats of rat- and insect-borne diseases, contaminated groundwater, and air pollution.

21. In a sanitary landfill, wastes are buried under earth.

22. In the environment, glass lasts longer than aluminum, which lasts longer than steel.

23. Reuse (rather than recycling) of materials involves the least expenditure of energy.

DISCUSSION

This chapter describes the resources of planet earth that are easily accessible to us. By far the most abundant atom is oxygen followed by silicon and hydrogen. Combinations of silicon and oxygen (sand is SiO_2 and silicates are SiO_4) form an important part of the materials used in our everyday life. Quartz, mica, and asbestos are different forms of pure silicates; ceramics (made by heating clays) are complex silicates; glass is sand mixed with various inorganic salts; cements and concretes are aluminum silicates mixed with limestone. Most metals are found combined with oxygen and have to be reduced to produce the pure metal. Copper was the first metal to be widely used followed by iron and steel. Although the most abundant metal, aluminum, is difficult to separate from oxygen, it is still cheaper to recycle aluminum than it is to process its ore. Since the earth neither gains nor loses appreciable mass and since mass is conserved, the effectiveness with which we manage our resources determines whether or not we have readily available resources or a nonpolluted environment. The other factor to consider in the availability of resources and quality of the environment is the world's population.

SELF-TEST

Multiple Choice

1. The mantle consists mostly of

    a. water   b. iron
    c. silicates   d. air

2. The most abundant element on the Earth's surface is

    a. oxygen   b. carbon
    c. hydrogen   d. silicon

3. The second most abundant element on the earth's surface is

    a. oxygen   b. carbon
    c. hydrogen   d. silicon

4. The three parts of the Earth's crust are the atmosphere, lithosphere, and

    a. hydrosphere   b. mantle
    c. stratosphere   d. hemisphere

5. Which is not a basic raw material for the production of iron?

    a. clay   b. coal
    c. iron ore   d. limestone

6. Quartz has the formula

    a. Si   b. SiO
    c. SiO_2 d. SiO_3

7. The principal raw materials for making glass are sand, sodium carbonate, and

    a. clay   b. coal
    c. iron ore   d. limestone

8. Which is not a silicate?

    a. asbestos   b. mica
    c. steel   d. glass

9. Clays are mostly

    a. pure silicates   b. aluminum silicates
    c. magnesium silicates   d. lead silicates

10. Raw materials used in cements are

    a. limestone and water   b. limestone and clay
    c. clay and water   d. clay and glass

11. The properties of steel can be varied by varying the proportion of

    a. oxygen   b. silica
    c. calcium oxide   d. carbon

12. Mica is easily cleaved into thin, transparent sheets. It is composed of SiO_4 tetrahedra in

    a. linear arrays   b. two-dimensional arrays
    c. three-dimensional arrays   d. irregular arrays

13. A metal is usually obtained from its ore by a process called

    a. corrosion   b. reduction
    c. casting   d. melting

14. Glass differs from crystalline silicates in that the particles that make up glass

    a. move more slowly   b. are smaller
    c. are randomly arranged   d. are not silicates

15. Asbestos, a fibrous mineral, is made of SiO_4 tetrahedra arranged in

    a. a three-dimensional array   b. sheets
    c. a linear array   d. random fashion

16. The earliest metal to be prepared from its ore is believed to be

    a. copper   b. silver
    c. gold   d. lead

17. Steel is an alloy of iron and

    a. copper   b. lead
    c. magnesium   d. carbon

18. The most abundant metal in the earth's crust is

    a. aluminum   b. iron
    c. calcium   d. magnesium

19. A thin layer of _____ protects aluminum from rusting

    a. carbon   b. iron
    c. aluminum oxide   d. iron oxide

20. Asbestos causes lung cancer. It is particularly deadly when it acts synergistically with

    a. benzene   b. sodium nitrite
    c. ozone   d. cigarette smoke

21. Low-grade ores are undesirable because they

    a. are difficult to mine
    b. cost more per ton of ore
    c. require more energy for concentration
    d. are found mainly in other countries

22. It takes _____ percent as much energy to make new aluminum cans from old aluminum cans as it does from aluminum ore.

    a. 5%   b. 10%
    c. 25%   d. 50%

23. We are in danger of exhausting our supply of lead because

    a. lead atoms are being destroyed
    b. lead metal is being converted to lead compounds
    c. lead ores are being converted to lead metal
    d. lead atoms are being scattered in the environment

24. Use of returnable bottles, averaging 12.5 uses before loss or destruction, involves what percentage as much energy per use as throwaway bottles?

    a. 10%   b. 50%
    c. 90%   d. 200%

25. Which is a threat from open dumps?

    a. air pollution   b. water pollution
    c. insect-borne disease   d. all of these are threats

26. Which lasts longest in the environment?

    a. glass   b. aluminum
    c. steel   d. paper

27. In a sanitary landfill, wastes are

    a. incinerated   b. composted
    c. buried under earth   d. aerobically digested

28.   Which process requires the least expenditure of energy?

        a.   reuse of an article (e.g., a glass bottle)
        b.   recycling an old article to make a new one
        c.   making a new article from virgin ores
        d.   making the article from a new kind of material

ANSWERS

1. c	8. c	15. c	22. a
2. a	9. b	16. a	23. d
3. d	10. b	17. d	24. a
4. a	11. d	18. a	25. d
5. a	12. b	19. c	26. a
6. c	13. b	20. d	27. c
7. d	14. c	21. c	28. a

Air

The Breath of Life

KEY TERMS

acid rain
allotropes
atmosphere
atmospheric inversion
catalytic converters
daughter isotopes

electrostatic precipitator
global warming
greenhouse effect
industrial smog
nitrogen cycle
oxygen cycle

ozone layer
particulate matter (PM)
photochemical smog
pollutant
Volatile Organic Compounds (VOC)
wet scrubbers

CHAPTER SUMMARY

12.1   The Atmosphere: Divisions and Composition
    A. The atmosphere is divided into layers.
        1. The troposphere is the layer next to the Earth's surface where nearly all life exists.
        2. The next layer is the stratosphere, where the Earth's protective ozone layer is located.
    B. Air is a mixture of gases. Dry air, by volume, is
        1. 78% nitrogen
        2. 21% oxygen
        3. 1% argon
    C. Variable components of air include
        1. 0% to 4% water vapor
        2. About 350 ppm carbon dioxide

12.2   The Nitrogen Cycle
    A. Plants need nitrogen as a nutrient, but cannot use N_2 molecules.
    B. Fixed nitrogen is atmospheric nitrogen combined with other elements.
    C. Some bacteria fix nitrogen; others convert it back to N_2. This establishes a nitrogen cycle.
    D. Lightning fixes nitrogen by causing it to combine with oxygen to make nitrogen oxides.
        1. The latter react with water to form nitric acid.
    E. Nitrogen is also fixed industrially to make nitrogen fertilizers (Chapter 13).

12.3   The Oxygen Cycle
    A. Our supply of oxygen is constantly replenished by green plants and consumed by animals and plants in the metabolism in foods.
    B. In the stratosphere, oxygen is formed by the action of ultraviolet rays on water molecules.
        1. Some oxygen is converted to ozone.
        2. Ozone shields us from harmful ultraviolet radiation.

C. An air pollution episode in Donora, Pennsylvania, in 1948 killed 17 people and made many others ill.
  1. The pollutants were sulfur dioxide and dust (probably fly ash).
  2. An atmospheric inversion—a warm, upper layer of air over a still, cooler lower layer—held the smog in the valley at Donora for 5 days.

12.4 Temperature Inversion
  A. Lower stagnant cold air is trapped by warmer air above it.
    1. Pollutants in the cold air are trapped near the ground sometimes for several days.

12.5 Natural Pollution
  A. Volcanoes spew ash and sulfur dioxide into the atmosphere.
  B. Dust storms add enormous amounts of particulate matter to the air.
  C. Swamps and marshes emit noxious gases.

12.6 The Air Our Ancestors Breathed
  A. Early people made fires, which added smoke to the atmosphere, and cleared land, which made dust storms worse.
  B. Rome was afflicted with stinking air and soot in A.D. 61.
  C. Heavy smoke made Nottingham unpleasant to the queen of England in 1257.
  D. The Industrial Revolution caused terrible pollution from burning coal in factory towns.
  E. The level of air pollution today is far greater than at other times in history.

12.7 Pollution Goes Global
  A. Huge urban areas today are afflicted with air pollution that drifts over rural areas.
    1. Smog drifts from Los Angeles to Colorado.
    2. Norway is afflicted with pollution from Germany and England.
    3. Most major cities around the world have suffered serious episodes of air pollution.
  B. A pollutant is a chemical in the wrong place at the wrong concentration.

12.8. Coal + Fire → Industrial Smog
  A. Industrial smog consists of smoke, fog, sulfur oxides, sulfuric acid, ash, and soot.
  B. An episode of air pollution in London in 1952 killed about 8,000 people.
  C. London smog is characterized by cool, damp weather with an atmospheric (thermal) inversion. The pollutants usually are produced by burning coal.
  D. The Chemistry of Industrial Smog
    1. When burned, the carbon in coal winds up as carbon dioxide, carbon monoxide, and soot (unburned carbon).
    2. Sulfur oxides
      a. The sulfur in coal is oxidized to sulfur dioxide (an acrid, choking gas), and then to sulfur trioxide, which reacts with water to form sulfuric acid.
    3. Particulate matter (solid and liquid particles of greater than molecular size) consists of minerals (fly ash) and soot.
      a. Visible particulate matter consists of dust and smoke.
      b. Invisible particulates are called aerosols (a dispersion of liquid particles in air).
      c. Unburned minerals in coal are known as clinkers.
        i. When clinkers are carried aloft in smokestacks, they are called fly ash.
    4. Health and Environmental Effects of Industrial Smog
      a. Sulfur dioxide and particulates such as ammonium sulfate act synergistically to cause far greater harm than either would alone.

b. Air pollution contributes to the development of respiratory diseases such as emphysema. (Cigarette smoking is a far larger contributor.)

c. Sulfur oxides and sulfuric acid also damage plants, causing crop losses.

E. What to Do about Industrial Smog

1. There are several ways to remove particulate matter from smokestack gases.

a. Electrostatic precipitators induce electric charges on the particles, which then are attracted to oppositely charged plates.

b. Bag filtration works much like a vacuum cleaner to clean the smokestack gases.

c. Cyclone separators cause the gas to spiral upward; particles hit the walls and settle out.

d. Wet scrubbers pass the stack gases through water, from which the particulates are removed.

2. Ash removed from stack gases can be used

a. To replace clay in making cement.

b. To make mineral wool for insulation.

c. The rest is stored in ponds and landfills.

3. It is difficult to remove sulfur oxides.

a. Sulfur can be removed from coal before burning

i. By flotation (Chapter 14).

ii. By gasification or liquefaction (Chapter 15).

b. Sulfur can be removed after burning by scrubbing the stack gases with a suspension of limestone or dolomite.

i. The calcium sulfite formed can be converted to the more useful calcium sulfate.

12.9 Photochemical Smog: Making Haze while the Sun Shines

A. Los Angeles smog (photochemical smog) occurs during dry, sunny weather.

B. The principal starting materials for photochemical smog are unburned hydrocarbons and nitrogen oxides from automobile exhausts.

C. Carbon Monoxide: The Quiet Killer

1. Carbon monoxide forms when a hydrocarbon burns in an insufficient amount of oxygen.

a. About 75% of the carbon monoxide that we dump into the atmosphere comes from automobile exhausts.

2. Carbon monoxide is odorless, tasteless, and invisible.

3. Carbon monoxide reacts with hemoglobin in the blood, hindering the transport of oxygen.

a. Chronic exposure to low levels of carbon monoxide adds stress to the cardiovascular system and may increase the chance of a heart attack.

b. Exposure to higher levels of carbon monoxide can cause drowsiness and death.

c. The best antidote is the administration of pure oxygen.

4. Nature somehow prevents the buildup of carbon monoxide; problems are local, not global.

D. Nitrogen Oxides: Brown Is the Color of Los Angeles Air

1. Any time combustion occurs in air, some of the nitrogen combines with oxygen to form nitric oxide. The higher the temperature, the more nitrogen oxide is formed.

2. Nitric oxide is slowly oxidized in air to amber-colored nitrogen dioxide.

$$2\,NO + O_2 \rightarrow 2\,NO_2$$

3. Photons from the sun split nitrogen dioxide into NO and reactive oxygen atoms.

a. The oxygen atoms produce a variety of irritating and toxic chemicals.

4. Nitrogen oxides produce smog and form nitric acid, which contributes to acid rain.

12.10 We Got the Lead Out
A. Tetraethyllead is an effective octane booster for gasoline, but the lead winds up in the environment.
B. Lead is toxic.
1. It affects the functioning of the blood, liver, kidneys, and brain.
2. It fouls the catalysts in catalytic converters.
C. The United States, Western Europe, and Japan have phased out the use of leaded gasoline.
D. To achieve the same octane rating in lead-free gasoline, more branched-chain and aromatic hydrocarbons are used.

12.11 Ozone: Protector and Pollutant
A. Ozone
1. In the stratosphere, ozone protects us from lethal ultraviolet radiation.
2. In the troposphere, it is a pollutant.
a. A pollutant is a chemical substance out of place in the environment.
B. Ozone is formed and destroyed in a cyclic process in the stratosphere. Levels fluctuate, but human activities may contribute to the destructive part of the cycle.
1. Ozone absorbs short wavelength ultraviolet rays.
2. During absorption, ozone is converted to an oxygen molecule and a reactive oxygen atom.
C. Ozone as an air pollutant
1. Ozone is a powerful oxidizing agent.
2. Ozone causes respiratory problems in humans and animals.
3. Ozone causes rubber to harden and crack.
4. Ozone causes extensive crop damage. especially to tomatoes and tobacco.
D. Chlorofluorocarbons and Other Threats to the Ozone Shield
1. Chlorofluorocarbons are insoluble in water and inert toward most substances. They persist in the environment for a long time.
2. Chlorofluorocarbons are broken down in the stratosphere by ultraviolet light to fragments including chlorine atoms.
3. These chlorine atoms catalyze the destruction of ozone.
4. For each 1% depletion of ozone, the U.S. National Research Council predicts a 2-5% increase in skin cancer.
a. A thinning or "hole" in the ozone layer has been found over the Arctic and Antarctica.
i. Evidence points to chlorofluorocarbons as the culprit.
5. Chlorofluorocarbons have been banned from aerosol spray cans in the United States, but are still used as refrigerants.
a. Chlorofluorocarbons were phased out completely in 1996.
6. Nuclear explosions in the stratosphere would produce nitrogen oxides. These also catalyze the destruction of ozone. Nuclear war might lead to an "ultraviolet summer."
a. Nuclear explosions near the Earth's surface might produce enough smoke and dust to block out sunlight leading to a "nuclear winter."
E. Volatile Organic Compounds (VOC)
1. Natural sources release hydrocarbons; only 15% of those in the atmosphere are there as the result of human activity.
2. In urban areas, the processing and use of gasoline contribute substantially to atmospheric hydrocarbons.
3. Hydrocarbons react with
a. Atomic oxygen or ozone to form aldehydes.
b. Oxygen and nitrogen dioxide to form peroxyacetyl nitrate (PAN).

4. Ozone, aldehydes, and PAN are irritants and contribute to much of the destruction wrought by photochemical smog.
　F.　What to Do about Photochemical Smog
　　　1. Hydrocarbon emissions have been reduced by new storage and dispensing systems.
　　　2. Catalytic converters on automobiles enhance the oxidation of hydrocarbons (and carbon monoxide) and reduce emissions of these two pollutants.
　　　　a.　Reducing nitrogen oxides is more difficult: NO must be reduced to N_2.
　　　　b.　With a catalyst, CO can be used to reduce NO to N_2.

12.12　Acid Rain: Air Pollution and Water Pollution
　A.　Sulfur oxides become sulfuric acid and nitrogen oxides become nitric acid. These acids lower the pH of rainwater.
　　　a.　Rain with a pH below 5.6 is called acid rain.
　　　b.　Normal rainwater is slightly acidic due to dissolved CO_2.
　B.　Good evidence indicates that acid rain comes from sulfur oxides and nitrogen oxides emitted from power plants, smelters, and automobiles.
　C.　Acid rain corrodes metals and destroys marble statuary.

12.13　The Inside Story: Indoor Air Pollution
　A.　Home: No Haven from Air Pollution
　　　1. Indoor air is often as bad as or worse than that outside.
　　　　a.　Gas ranges and kerosene heaters produce nitrogen oxides.
　　　　b.　Formaldehyde is released from foamed insulation, fiberboard, and particle board.
　B.　Cigarettes and Second Hand Smoke
　　　1. Cigarette smoke is the most prevalent indoor air pollutant.
　　　2. Forty carcinogens have been found in cigarette smoke.
　　　3. Carbon monoxide levels often exceed standards for ambient air.
　　　4. Nonsmokers are also exposed to tars, nicotine, and allergy-triggering substances in cigarette smoke.
　C.　Radon and Her Dirty Daughters
　　　1. Radon is a colorless, odorless, tasteless, chemically unreactive, radioactive gas found in rocks (granite and shale) and minerals.
　　　　a.　Radon decays to polonium-218 (daughter isotope), lead-214, and bismuth-214, which are trapped in the lungs and damage tissue.
　　　　b.　Trapped inside well-insulated houses, radon levels build up and exceed EPA limits.

12.14　Who Pollutes? How Much?
　A.　Automobiles produce one half of all air pollutants, including
　　　1. 80% of carbon monoxide emissions.
　　　2. 40% of hydrocarbon emissions.
　　　3. 40% of nitrogen oxide emissions.
　B.　Power plants produce
　　　1. 80% of sulfur oxide emissions.
　　　2. 40% of particulate emissions.
　　　3. 55% of nitrogen oxide emissions.
　C.　Other industries contribute
　　　1. 45% of particulate emissions.
　　　2. 15% of sulfur oxide emissions.
　D.　Air quality overall has improved over the last two decades.
　　　1. Lead emissions are down 96%.
　　　2. Particle emissions are down 61%—especially true in urban areas.

3. Carbon monoxide levels are down 38%, but many cities still exceed health standards.
4. Sulfur oxide emissions are down 28%.
5. Nitrogen oxide levels have increased 8%.

12.15 Carbon Dioxide and Global Warming
A. Nearly all combustible processes yield carbon dioxide as one of their products.
B. Carbon dioxide levels have increased 18% in this century.
C. Greenhouse Effect: Planet with a Fever
1. Carbon dioxide and other gases let sunlight in, but hinder the release of infrared radiation to space, causing the Earth to get warmer (the greenhouse effect).
   a. Human activities add 25 billion tons of CO_2 to the atmosphere each year.
   b. Plants, soil, and oceans remove 15 billion tons.
2. Methane and other trace gases contribute to the greenhouse effect; methane concentrations are rising.
   a. Methane is 20 to 30 times more efficient at trapping heat as CO_2.
   b. Chlorofluorocarbons are 20,000 times more efficient at trapping heat than CO_2.
D. Many scientists believe that global warming is melting ice caps and causing oceans to rise.

12.16 The Ultimate Pollutant: Heat
A. Every energy conversion involves the production of waste heat.
B. Heat may be the ultimate pollutant, eventually changing the climate of the Earth.

12.17 Paying the Price
A. Air pollution costs the United States billions of dollars per year.
B. Costs increase rapidly as we try to remove larger percentages of pollutants.

CHAPTER OBJECTIVES

(You should know that...)

1. The layer of the atmosphere nearest the Earth, the layer in which we live, is the troposphere.

2. The ozone layer is found in the stratosphere.

3. The atmosphere is roughly four-fifths nitrogen and one-fifth oxygen.

4. Water vapor and carbon dioxide are variable components of the Earth's atmosphere.

5. Nitrogen must be fixed (combined with other elements) before it can be used by plants.

6. Nitric oxide (NO) is formed in thunderstorms, automobile engines, power plants, and industrial plants.

7. An atmospheric (thermal) inversion consists of a warm layer of air over a cool, stagnant layer.

8. Volcanoes and dust storms are natural sources of pollutants.

9. Industrial smog is characterized by cool, damp weather, and by sulfur oxides and particulate matter.

10. The most dangerous component of industrial smog is sulfur dioxide.

11. The burning of coal is responsible for most sulfur dioxide emissions.

12. The most serious disadvantage of low-sulfur coal is that it usually has a lower heating value than high-sulfur coal.

13. Sulfur dioxide and particulates act synergistically in causing respiratory problems.

14. Carbon monoxide binds to hemoglobin, blocking the transport of oxygen.

15. Carbon monoxide is the air pollutant produced in greatest quantity.

16. Electrostatic precipitators, bag filters, cyclone separators, and wet scrubbers are used to remove particulate matter from smokestack gases.

17. The essential ingredient for the production of photochemical smog is nitrogen dioxide.

18. The ozone layer is threatened by chlorofluorocarbons (Freons) and nuclear explosions.

19. The most significant air polluter, in terms of amount of gases emitted, is the internal combustion (automobile) engine.

20. Autos are a major source of hydrocarbons, carbon monoxide, and nitrogen oxides.

21. Ozone causes rubber to harden and crack.

22. Nitrogen oxide emissions from an internal combustion engine can be lowered by increasing the fuel-to-air ratio or by operating at a lower temperature. (The carbon monoxide and hydrocarbon emissions would be increased, however.)

23. Increasing the carbon dioxide content of the atmosphere may increase the temperature of the Earth's atmosphere (greenhouse effect).

24. Rain in areas downwind from industrialized regions is often quite acidic.

25. Acid rain is that with a pH below 5.6.

26. Normal rain is slightly acidic due to dissolved CO_2.

27. The greenhouse effect can cause the Earth to get warmer and the ice caps to melt, resulting in the oceans rising.

EXAMPLE PROBLEMS

1. There is 0.36 L of carbon dioxide in 1000 L of air. What is the concentration of CO_2 in air in parts per million (by volume)? Parts per million means parts in a million number of parts. The "parts" may be liters or any other convenient unit. To find out the answer, multiply both the numerator and denominator of the ratio

$$\frac{0.36 \text{ L } CO_2}{1000 \text{ L air}}$$

by 1000 to get 1 million L of air.

$$\frac{.36 \text{ L } CO_2}{1000 \text{ L air}} \times \frac{1000}{1000} = \frac{360 \text{ L } CO_2}{1,000,000 \text{ L air}}$$

We now have 360 L CO_2 per 1,000,000 L air, or 360 parts CO_2 per million parts air, or simply 360 ppm CO_2.

2. What mass of particulate matter would be inhaled each day by a person breathing 20,000 L of city air containing 230 $\mu g/m^3$ of particulate matter? (1 m^3 = 1000 L)

$$x \mu g \text{ particulate matter} = \frac{230 \text{ }\mu g}{1 \text{ } m^3} \times \frac{1 \text{ } m^3}{1000 \text{ L}} \times 20,000 \text{ L}$$

$$= \frac{(230)(1)(20,000)}{(1)(1000)}$$

$$X = 4600 \text{ }\mu g$$

ADDITIONAL PROBLEMS

1. At present, the atmosphere contains about 2.5 quadrillion kilograms (2.5×10^{15}kg) of carbon dioxide. By burning fossil carbon, we add about 22 trillion kilograms (22×10^{12}kg) of CO_2 to the atmosphere each year. If half of this CO_2 remains in the atmosphere, how many years will it take, at the present rate, to double the amount of CO_2 in the atmosphere?

2. A supersonic transport (SST) burns 60,000 kg of fuel per hour. What weight of carbon dioxide and of water vapor will be produced? A (representative) equation is

$$C_{15}H_{32} + 46 \text{ } O_2 \rightarrow 15 \text{ } CO_2 + 16 \text{ } H_2O$$

3. In 10,000 L of air there are 3 L of argon. What is the concentration of argon in parts per million (by volume)?

DISCUSSION

While nitrogen is 80% of the atmosphere, it needs to be combined with other elements (fixed) in order to be useful. The Earth puts out a number of pollutants via volcanoes, dust storms, swamps and marshes. However, since human beings started living in cities, man-made pollution has been a serious problem. Burning coal produces Industrial (London) smog which is a combination of smoke, fog, sulfur oxides, and particulate matter. To reduce Industrial smog requires the removal of sulfur from the fuel and smoke that produces particulates. Photochemical (Los Angeles) smog is produced by automobiles and contains carbon monoxide, nitrogen oxides, and unburned hydrocarbons. The best way to reduce photochemical smog is through the use of catalytic converters. Ozone is a pollutant at ground level but needed as a protective screen against UV rays in the upper atmosphere. CFCs threaten to diminish this protective layer of ozone. Acid rain occurs when sulfur oxides and nitrogen oxides combine with water to lower its pH. The United States successfully removed lead as an air pollutant by banning tetraethyllead as a gasoline additive. Tighter insulation has increased the level of indoor air pollution. Indoor air pollutants include nitrogen oxides from gas ranges, cigarette smoke, and radon. Table 12.2 in the text lists the major sources of each type of air pollutant. Carbon dioxide, produced by burning coal or hydrocarbons may contribute to global warming. "Waste" heat produced as a by-product during the conversion of one type of energy to another may contribute to global warming in the future.

SELF-TEST

Multiple Choice

1. Most humans live in the

    a. mesosphere
    c. troposphere

    b. stratosphere
    d. western hemisphere

2. What fraction of the air is nitrogen?

    a. 1/5
    c. 1/2

    b. 1/4
    d. 4/5

3. "Fixed" nitrogen is

    a. pure nitrogen
    c. nitrogen combined with another atom

    b. diatomic nitrogen
    d. all of the above

4. Which of the following components of the atmosphere vary in concentration?

    a. oxygen
    c. argon

    b. nitrogen
    d. water vapor

5. Which are sources of natural pollution?

    a. volcanoes
    c. swamp and marsh gases

    b. dust storms
    d. all of the above

6. The most significant air polluter, in terms of total amounts of gases emitted each year in the United States, is

   a. motor vehicles
   c. industry
            b. electric power plants
            d. space heating

7. Temperature inversion

   a. can trap pollutants near the ground
   c. is warm air trapped under cold
            b. is a natural phenomenon
            d. is all of the above

8. Which condition is associated with Industrial rather than photochemical smog?

   a. sunny, dry weather
   c. operation of automobiles
            b. combustion of high-sulfur coal
            d. high ozone levels

9. Air pollution

   a. started after World War II
   b. started after the Revolutionary War
   c. started during the Dark Ages
   d. started when people started building cities

10. Which of these compounds is one of the primary pollutants associated with Industrial-type smog?

   a. NO
   c. SO_2
            b. O_3
            d. PAN

11. Partial but incomplete combustion of carbon yields

   a. CO
   c. CH_4
            b. CO_2
            d. NO

12. Of the following toxic gases, which is colorless, odorless, and tasteless?

   a. NO_2
   c. NH_3
            b. H_2S
            d. CO

13. Which pollutant is probably responsible for the observed lowering of the Earth's average temperature?

   a. O_3
   c. particulates
            b. CO_2
            d. methane

14. Which is a combination that exhibits a harmful synergistic effect?

   a. sulfur dioxide and particulate matter as pollutants
   b. oxygen and nitrogen as a breathing mixture
   c. ozone and aerosol propellants in the upper atmosphere
   d. radon and helium

15. Nitrogen fixation is accomplished by

   a. some bacteria
   c. automobile engines
   b. lightning
   d. all of these

16. The air pollutant that successfully competes with oxygen for bonding sites in hemoglobin molecules of blood is

   a. SO_2
   c. CO
   b. CO_2
   d. O_3

17. An increase in the amount of particulate matter may result in a cooler environment because

   a. the Earth's reflectivity is decreased
   b. the greenhouse effect of CO_2 is decreased
   c. particles reflect away solar radiation
   d. particles serve as nuclei for snow to form

18. Ozone is good in the _____ but a pollutant in the _____

   a. stratosphere   troposphere
   c. troposphere   stratosphere
   b. troposphere   mesosphere
   d. mesosphere   troposphere

19. Ozone is a poison because it

   a. is a heavy metal poison
   c. is a strong oxidizing agent
   b. bonds to iron in hemoglobin
   d. reacts with carbon dioxide

20. Scientists are opposed to chlorofluorocarbons as aerosol propellants because they

   a. are highly toxic
   b. destroy carbon monoxide in the lower atmosphere
   c. hasten decomposition of the ozone layer in the upper atmosphere
   d. don't last very long

21. An essential ingredient in the production of photochemical smog is

   a. carbon monoxide
   c. sulfur dioxide
   b. carbon dioxide
   d. nitrogen dioxide

22. The most dangerous component of Industrial smog is

   a. PAN
   c. nitric oxide
   b. soot
   d. sulfur dioxide

23. A major source of hydrocarbons is

   a. industry
   c. coal burning power plants
   b. nuclear power plants
   d. automobiles

24. The burning of coal is most responsible for which of the following air pollutants?

    a. CO   b. SO_2
    c. particulate matter   d. hydrocarbons

25. Catalytic converters are used to reduce

    a. ozone   b. hydrocarbons
    c. water   d. CFCs

26. Which of the following gases is likely to be a pollutant emitted by the internal combustion engine regardless of the fuel used?

    a. benzene   b. SO_2
    c. NO   d. particulate matter

27. Which device is not used to remove particulates from smokestack gases?

    a. bag filter   b. cyclone precipitator
    c. electrostatic precipitator   d. photovoltaic cell

28. Nuclear explosions in the stratosphere might lead to

    a. an increase in ozone levels   b. a decrease in NO levels
    c. a nuclear winter   d. an ultraviolet summer

29. The greenhouse effect results from an increase in the concentration of one of the following substances in the atmosphere. Which one?

    a. CO_2 b. NO_2
    c. O_3 d. particulates

30. Which of the following actions would result in an increase in the temperature of the Earth-atmosphere system?

    a. increase the distance from the sun
    b. remove water vapor from the atmosphere
    c. increase the cloud cover of the Earth
    d. increase the carbon dioxide content of the atmosphere

31. Which pollutant contributes significantly to the formation of acid rain?

    a. SO_2 b. NH_3
    c. CO   d. hydrocarbons

32. Because of the greenhouse effect, the temperature of the Earth is expected to

    a. increase   b. decrease
    c. fluctuate wildly   d. remain the same

33.   In which location is ozone considered to be beneficial to life on Earth?

   a. in the stratosphere
   b. at ground level
   c. in the mesosphere
   d. Ozone is extremely toxic and is never considered to be beneficial to life.

34.   Radon is caused by

   a. naturally occurring rocks
   b. nuclear power plants
   c. nuclear submarines
   d. nuclear bomb testing in the atmosphere

Matching

Match the pollutant with its principal source.   Note: Sources may contribute to more than one pollutant.

b 35. Carbon monoxide   a. industry
b 36. Hydrocarbons   b. motor vehicles
b 37. Lead   c. power plants
c 38. Nitrogen oxides   d. secondary pollutant
d 39. Ozone
a 40. Particulate matter
c 41. Sulfur oxides

ANSWERS

Additional Problems
1. 454.5 years.
2. 187,000 kg carbon dioxide; 82,000 kg water
3. 300 ppm

Self-Test

1. c	8. b	15. d	22. d	29. a	36. b
2. d	9. d	16. c	23. d	30. d	37. b
3. c	10. c	17. c	24. b	31. a	38. c
4. d	11. a	18. a	25. b	32. a	39. d
5. d	12. d	19. c	26. c	33. a	40. a
6. a	13. c	20. c	27. d	34. a	41. c
7. d	14. a	21. d	28. d	35. b	

Water

Rivers of Life; Seas of Sorrow

KEY TERMS

activated sludge method	dissolved oxygen (DO)	primary sewage treatment
advanced treatment	eutrophication	reverse osmosis
aerobic oxidation	hard water	secondary sewage treatment
anaerobic decay	heat capacity	specific heat
biochemical oxygen demand (BOD)	heat vaporization	

CHAPTER SUMMARY

13.1   Water: Some Unusual Properties
    A.   Water is the only common liquid on the surface of planet Earth.
    B.   Solid water (ice) is less dense than liquid water.
    C.   Water has a higher density than other familiar liquids.
    D.   Water has an unusually high heat capacity. (The amount of heat a substance can absorb before its temperature rises 1°C).
    E.   Water has an unusually high heat of vaporization. (Amount of heat required to evaporate a small amount of water).
    F.   The properties of water are explained by its structure.
        1.   Liquid water is strongly associated through hydrogen bonding, but the association is random.
        2.   In ice, molecules have a more ordered arrangement, with large hexagonal holes.

13.2   Water, Water, Everywhere
    A.   Ninety-eight percent of all water on the surface of the Earth is seawater.
    B.   Water is polar; it tends to dissolve ionic substances.
    C.   Water dissolves minerals that are carried to the sea by rivers. Water evaporates from the oceans, leaving the salts behind.
        1.   The oceans are ever so slowly becoming more salty.

13.3.   The Water Cycle
    A.   Natural Water Isn't All H_2O.
        1.   Rainwater contains dust, dissolved gases (carbon dioxide, oxygen, nitrogen), and in thunderstorms, nitric acid.

2. Groundwater contains dissolved ions.
    a. The principal positive ions are sodium, potassium, calcium, magnesium, and sometimes iron.
        i. Calcium, magnesium, and iron ions are responsible for hard water.
    b. The principal negative ions are sulfate, bicarbonate, and chloride.
B. Water evaporates from the oceans, lakes, and land; salts are left behind.
C. Water returns to Earth as fresh water (rain and snow).
D. Water moving through the ground is purified as impurities are trapped by sand, gravel, and clay.
E. Some Biblical Chemistry
    1. Moses purified bitter water by throwing a tree into it.
        a. The oxidized cellulose of the tree may have neutralized the alkali in the water.

13.4 Biological Contamination: The Need for Clean Water
A. Water-borne diseases (cholera, typhoid fever, dysentery) plagued the entire world until early this century.
B. Chemically treated community water supplies have almost eliminated water-borne diseases in industrialized nations.
C. Water-borne diseases are common today in much of the world.
    1. About 80% of all illnesses are caused by contaminated water.
    2. People with such diseases fill half of the world's hospital beds.
    3. They die at a rate of 25,000 per day.
    4. Fewer than 10% of the Earth's people have adequate amounts of clean water.
D. Biological contamination is still a threat in the United States.
    1. Perhaps 30 million people are at risk because of bacterial contamination.
    2. Hepatitis (a viral disease) is sometimes spread through drinking water.
    3. Biological contamination may have peaked in the 1960s; much improvement has been made since then, but biological contamination remains the greatest threat to our water supply.

E. Sewage: Some Chemistry and Biology
    1. Pathogenic microorganisms cause disease.
    2. Breakdown of sewage depletes dissolved oxygen and adds plant nutrients to water.
        a. Degradation can be aerobic (with air) or anaerobic (without air).
        b. Biochemical oxygen demand (BOD) is a measure of the amount of oxygen required to degrade the organic material in water.
    3. With adequate dissolved oxygen, aerobic bacteria degrade organic wastes to carbon dioxide, water, nitrates, phosphates, and sulfates.
        a. Nitrates and phosphates fertilize the water, stimulating algae growth.
        b. When the algae die, their decomposition requires oxygen (increases the BOD), a process called eutrophication.
    4. With the oxygen depleted, anaerobic bacteria flourish. Anaerobic decay leads to foul-smelling sulfur compounds, ammonia and amines.

13.5 Ecological Cycles
A. In a simplified cycle, fish produce wastes, bacteria break down the waste to inorganic materials that serve as nutrients for algae, and fish eat the algae.
B. Humans interfere with the cycle in several ways.
    1. We increase the load of wastes by dumping sewage into the water, increasing the BOD.
    2. We introduce nutrients by runoff of fertilizer and phosphates from detergents leading to algae blooms.

3. We introduce pesticides, detergents, toxic metals, plastics, radioisotopes, and other harmful materials into the water.
   C. Chemists can monitor the BOD and determine the level of harmful substances.

13.6   Chemical Contamination: From Farm, Factory, and Home
   A. Industries have dumped chemical wastes into waterways.
   B. Fertilizers and pesticides from farms and lawns enter the waters.
   C. Transportation results in spills of oil and chemicals.
   D. Household chemicals are dumped down drains.
   E. Chemistry is required to identify the often invisible materials in water.

13.7   Groundwater Contamination: Tainted Tap Water
   A. Half of the people in the United States depend upon groundwater for drinking water.
      1. Toxic substances have been found in many community and private wells. Poisons in the wells include
         a. Industrial wastes.
         b. Pesticides (aldicarb in potato-growing areas).
   B. Nitrates
      1. In many agricultural areas, well water is contaminated with nitrates (from fertilizer).
      2. Levels greater than 10 ppm are dangerous to infants, causing ethemoglobinemia (blue baby syndrome).
      3. Only expensive advanced treatment can remove highly soluble nitrates from water.
   C. Volatile organic chemicals in groundwater.
      1. Volatile organic chemicals are used in homes and factories as solvents, cleaners, and fuels.
         a. Common volatile organic contaminants (VOC) are hydrocarbons (benzene and toluene) and chlorinated hydrocarbons (carbon tetrachloride, chloroform, methylene chloride, and trichloroethylene).
         b. Except for toluene, all are suspected carcinogens.
      2. Only traces of commonly used volatile organic chemicals dissolve in water.
   D. Leaking Underground Storage Tanks: UST.
      1. Underground storage tanks at service stations often rust and leak gasoline into the surrounding soil and water.
   E. Trace toxics and public perception.
      1. Amounts of contaminants are often in the parts per million, parts per billion, or parts per trillion range.
   F. Calculations of parts per million and parts per billion
      1. 1 ppm = 1 mg solute/ Liter of solution
         Change 4.2 mg Cl⁻/L to ppm and ppb
         Assume Density of water = 1g/mL so we have 4.2 mgCl⁻/1000g
         (4.2 mg Cl⁻/1000 g water) x (1 g water/1000mg water) =
         4.2 mg Cl⁻/1,000,000 mg water or 4.2 ppm
         1000 ppb = 1ppm so   (4.2 ppm) x (1000 ppb/1 ppm) = 4200 ppb

13.8   Acid Waters: Dead Lakes
   A. Acids enter our waters through acid precipitation (acid rain, snow, fog or dry deposition) and through drainage from mines.
   B. Acids cause toxic ions to be released from rocks and minerals.
      1. Example: Acid rain releases aluminum ions from clay. These ions are especially toxic to young fish.
      2. Alzheimer's victims show high levels of aluminum ions in defective brain cells.

C. Lakes in limestone areas are not seriously harmed because the limestone neutralizes the acids.
D. There are several possible solutions to the problem of acidic waters.
   1. Neutralize acidic lakes with pulverized limestone.
   2. Remove sulfur before coal is burned.
   3. Scrub sulfur oxides from smokestack gases.

13.9   Industrial Water Pollution
A. Most chemical industries are in compliance with the U.S. Water Pollution Control Laws.
B. Automobiles and water pollution
   1. Chrome plating dumps cyanide and chromate ions into the water. Processes for removing these pollutants include:
      a. Cyanide ions can be converted by chlorine to bicarbonate ions and nitrogen gas.
      b. Chromate salts can be reduced by sulfur dioxide to Cr^{3+} ions, which can be removed by precipitation in basic solution.
C. Water pollutants from other industries
   1. Wastes from textile industries and from meat and other food-processing plants can be treated by sewage plants.
   2. Oil refineries and chemical plants have made substantial reductions in wastes produced.

13.10   From Waste Water to Drinking Water
A. Primary sewage treatment plants remove some solids as sludge.
   1. The effluent water has a high Biochemical Oxygen Demand (BOD).
B. Secondary sewage treatment plants pass the effluent through sand and gravel filters.
   1. This provides some aeration, aiding the action of aerobic bacteria.
C. The activated sludge method combines primary and secondary treatment.
   1. Sewage is aerated with large blowers forming large porous layers.
   2. Part of the sludge is recycled.
D. Sewage effluent is chlorinated to kill pathogenic bacteria.
   1. Excess chlorine provides residual protection.
   2. Chlorine is not effective against some viruses.
   3. Chlorine reacts with organic compounds to form chlorinated hydrocarbons, some of which are carcinogens.
E. Some European cities use ozone rather than chlorine to kill pathogens.
   1. Ozone is more effective against viruses.
   2. Ozone yields oxidated, not chlorinated, hydrocarbons.
   3. Ozone imparts no "chemical" taste to the water.
   4. Unlike chlorine, ozone provides no residual protection.
F. Advanced (tertiary) treatment
   1. Charcoal filtration removes dissolved organic compounds.
   2. Reverse osmosis (pressure filtration) removes most ions.
G. A Drop to Drink
   1. Water from reservoirs, rivers, and lakes must be treated to make it safe for drinking.
      a. Water is treated with slaked lime and aluminum sulfate.
         i. The aluminum hydroxide formed carries down dirt and bacteria.
      b. The water is then filtered through sand and gravel.
      c. Sometimes the water is aerated to remove odors and improve the taste.
      d. Sometimes the water is filtered through charcoal to remove colored and odorous compounds.
      e. Finally, the water is chlorinated to kill harmful bacteria.
   2. Fluorides
      a. Fluorides are poisonous in moderate to high concentrations.

b. In concentrations of 0.7 to 1.0 ppm fluorides in drinking water lead to a reduction in the incidence of dental cavities (caries).
    i. Fluorides strengthen tooth enamel by converting hydroxyapatite to fluorapatite.
    ii. Excessive fluorides can cause a mottling of tooth enamel and interfere with calcium metabolism, kidney action, and thyroid function.

13.11   Back to the Soil: An Alternative Solution
    A. Nutrients from wastes could be returned to the soil instead of dumped into water.
      1. In many societies, human and animal wastes are returned directly to the soil.
      2. Sludge is used as fertilizer in some parts of the United States.
      3. Toilets are available that use no water; they compost wastes.
    B. Some disadvantages of using sludge as fertilizer.
      1. Pathogenic organisms may survive and spread disease.
      2. The sludge is often contaminated with toxic metal ions.

13.12   You're the Solution to Water Pollution
    A. Conserve.
    B. Be prepared to pay the bill for effective water treatment.
    C. Only about 1.5 L of drinking water per day is needed for drinking.
    D. Average daily direct use in the United States:
      1. 7 L/day for drinking and cooking.
      2. 120 L/ day for bathing, washing clothes and dishes, etc.
      3. 80 L/day for flushing the toilet.
      4. 85 L/day for swimming pools and lawns.
    E. Indirect uses (in agriculture and industry):
      1. 1 kilogram of vegetables requires 800 L of water.
      2. One steak requires 13,000 L of water.
      3. Recreational waters (boating and swimming) must also be free of disease-causing organisms.

CHAPTER OBJECTIVES

(You should know that...)

1. Because water molecules are associated through hydrogen bonding, water has high boiling and high melting temperatures, a high heat of vaporization, and a high heat capacity.

2. Specific heat is the amount of heat required to raise the temperature of 1g of a substance by 1°C.

3. In 1900, the most serious threat to health from water pollution in the United States was bacteria. For much of the world, that is still true today.

4. Rainwater, even in remote areas, usually contains dust, dissolved oxygen, and (if there is lightning) nitric acid.

5. Waterfalls and rapids add more dissolved oxygen to streams.

6. Water is the only common liquid on the Earth.

7. Solid water (ice) is less dense than liquid water.

8. Water has a high heat capacity.

9. Water has a higher density than most other familiar liquids.

10. Water has an exceptionally high heat of vaporization.

11. Water tends to dissolve polar and ionic substances.

12. Domestic sewage is the major source of water pollution.

13. A sewage plant with secondary treatment removes most oxygen-consuming wastes.

14. Oil, composed of nonpolar molecules, will not dissolve in water.

15. Solubility of gases in water (important for aquatic life) decreases with an increase in temperature (and increases with an increase in pressure).

16. High Biochemical Oxygen Demand (BOD) is caused by dumping organic wastes (sewage, paper-mill effluent, slaughterhouse wastes, cannery wastes, etc.) into water.

17. Groundwater has been contaminated by dumps, agricultural runoff of pesticides and fertilizers, and chemical solvents such as hydrocarbons (benzene and toluene) and chlorinated hydrocarbons (carbon tetrachloride, chloroform, methylene chloride, and trichloroethylene).

18. UST is an acronym for leaking Underground Storage Tanks, a source of gasoline in groundwater.

19. Drinking water is fluoridated to reduce the incidence of tooth decay.

20. Drinking water is chlorinated to kill pathogenic (disease-causing) bacteria. Chlorine kills bacteria, but not viruses. It converts dissolved organic compounds into chlorinated hydrocarbons, some of which are known carcinogens.

21. Some form of tertiary treatment is required to remove nitrates and phosphates from sewage.

22. BOD is a measure of the biodegradable material in water. (Bacteria use oxygen to degrade wastes.)

23. Decomposition of organic matter in water is usually aerobic.

24. Anaerobic decomposition releases foul-smelling hydrogen sulfide.

25. Nitrates and phosphates are plant nutrients. They, along with carbon dioxide and water, are the ultimate aerobic degradation products of sewage.

26. Strip mining results in acid runoff in streams.

(In addition, your instructor may require you to know what type of sewage treatment is used in your community, where your water supply comes from and how it is treated, etc.)

EXAMPLE PROBLEMS

1.   The U.S. Food and Drug Administration (FDA) limit for mercury in food is 0.5 ppm. On a seafood diet, a person might consume 340 g (12 oz) of tuna per day. How much mercury would the person get each day if the tuna contained the maximum of mercury? (A mg is 1×10^{-6} g).

$$340 \text{ g tuna} \quad X \quad \frac{.5 \text{ g Hg}}{1,000,000 \text{ g tuna}} \quad == \quad 0.00017 \text{ g}$$

$$0.00017 \text{ g} \quad X \quad \frac{1,000,000 \text{ microgram}}{1 \text{ g}} \quad == \quad 170 \text{ microgram}$$

(The lowest level at which toxicity symptoms have been observed is about 300 mg/day.)

2.   Every 3 parts of organic matter in water requires about 8 parts of oxygen for its degradation. How much organic matter is needed to deplete the 10 ppm of oxygen in (a) 1 L (which weighs 1 kg) of water? (b) A small lake containing 1,000,000,000 kg of water?

a. $$1 \text{ kg water} \quad X \quad \frac{10 \text{ g oxygen}}{1,000,000 \text{ g water}} \quad X \quad \frac{3 \text{ g organix}}{8 \text{ g oxygen}} \quad == \quad \begin{array}{l} 0.000004 \text{ kg oxygen} \\ \text{or 4 mg organic matter} \end{array}$$

b. $$1,000,000,000 \text{ kg } H_2O \quad X \frac{10 \text{ g oxygen}}{1,000,000 \text{ g water}} \quad X \quad \frac{3 \text{ g organix}}{8 \text{ g oxygen}} \quad == \quad 4000 \text{ kg organic matter}$$

ADDITIONAL PROBLEMS

1.   Sewage discharged by each person each day in the United States consumes, on the average, 60 g of oxygen. How many kilograms of water, at 10 ppm O_2, are depleted daily by the raw sewage from a city of 100,000 people?

2.   Arizona has estimated recoverable groundwater of 8.7×10^{14} kg. At present, the water is being used at a net rate of 5×10^{12} kg/year. At this rate, how long will the water last?

DISCUSSION

Water has some very unusual properties that make life possible. Although three-fourths of the earth is covered with water, only 1% of the world's water is fresh water. The water cycle (evaporation and condensation) cleans water. Until 100 years ago, the leading cause of death was from contamination of water by human waste. Even now 80% of the world's sickness is caused by contaminated water. Biochemical oxygen demand (BOD) measures the amount of organic material dissolved. Groundwater can be contaminated by volatile organic chemicals and leakage from underground storage tanks (UST). Recent abilities to detect lower and lower concentrations have led to overdramatization of pollution problems. Acid rain and runoff from abandoned mines can cause lakes to become acidic. Industry uses a lot of water in its processes with a lot of chances to pollute. Waste water can be treated by settling (primary), settling and filtering (secondary), and settling, filtering, and advanced treatment (tertiary) methods. Drinking water is often treated with aeration and chlorination. Fluoridation is used to reduce tooth decay but the concentrations must be carefully monitored. In rural areas, nitrates from farming operations are a problem.

SELF-TEST

Multiple Choice

1. _____ is the only substance to exist in large amounts in all three physical states.

   a. mercury b. methane
   c. crude oil d. water

2. What percent of the world's water is in the oceans?

   a. 2% b. 25%
   c. 50% d. 98%

3. What percent of the world's water is available as fresh water?

   a. 1% b. 5%
   c. 10% d. 20%

4. Which property of water is regarded as unusual?

   a. its specific heat
   b. relative densities of solid and liquid
   c. heat of vaporization
   d. all of these

5. Because water molecules are associated through hydrogen bonding, water has a high

   a. temperature b. density
   c. boiling point d. solubility

6. The only common liquid on the Earth is

   a. mercury b. water
   c. petroleum d. oxygen

7. Which is not a special property of water?

    a. solid less dense than liquid
    b. high heat capacity
    c. low density compared to that of most other liquids
    d. high heat of vaporization

8. Large hexagonal holes in the arrangement of water molecules in ice account for the fact that water

    a. dissolves ionic compounds   b. is easily polluted
    c. has a high heat capacity   d. expands upon freezing

9. In the water cycle, which process removes salts?

    a. raining
    b. water stored in lakes
    c. water condensing from clouds
    d. water evaporating from the surface

10. What percent of the world's sickness is caused by contaminated water?

    a. 10%   b. 20%
    c. 40%   d. 80%

11. What kind of oil molecule would account for the fact that oil slicks do not dissolve in water?

    a. nonpolar   b. polar
    c. subpolar   d. arctic

12. Hard water contains

    a. ice
    b. ions such as sodium, potassium, and chloride
    c. ions such as calcium, magnesium, and iron
    d. ions such as hydronium and hydroxide

13. The main source of water pollution is
    a. industry   b. domestic sewage
    c. chemical plants   d. fertilizer runoff

14. The source of volatile organic chemicals in groundwater is most likely to be

    a. gasoline stations   b. dry cleaners
    c. old dumps   d. industry

15. Which of the following analyses provides a measure of the biodegradable material present in waste water?

    a. COD   b. BOD
    c. pH   d. hardness

16. Which of the following is a product of the anaerobic decomposition of sewage?

    a. mercury   b. lead
    c. H_2S d. SO_2

17. Which of the following water pollutants was the most serious threat to human health in 1900?

    a. mercury   b. phosphates
    c. nitrates   d. bacteria

18. The major source of acid rain is

    a. automobiles   b. carbonated drinks
    c. burning coal   d. burning natural gas

19. The most commonly used agent for killing bacteria in treated waste water is

    a. chlorine   b. ozone
    c. gamma radiation   d. phosphates

20. Which of the following is not true?

    a. solubility of gases in water increases with increasing temperature
    b. solubility of gases in water increases with increasing pressure
    c. solubility of gases in water is important for fish populations
    d. solubility of both oxygen and nitrogen gases increases with pressure

21. Which material requires the most water to produce?

    a. paper   b. steel
    c. aluminum   d. synthetic rubber

22. Which company's industrial sewage would have a relatively high BOD?

    a. Sidewalk Cement Company
    b. Desecration Strip Mining Company
    c. Kosher Slaughterhouse Company
    d. Bethlehem Steel Works

23. Water treatment that treats water by settling is

    a. primary   b. secondary
    c. tertiary   d. quaternary

24. Water treatment that treats water by settling and by filtration and aeration is

    a. primary   b. secondary
    c. tertiary   d. quaternary

25. Which of the following could be used in a tertiary water treatment?

   a. ion exchange
   b. carbon bed with regeneration
   c. reverse osmosis
   d. all of the above

26. In the activated sludge process for treating waste water, the most important step is

   a. chemical precipitation
   b. biological decomposition
   c. chlorination
   d. reverse osmosis

27. Activated charcoal is used in water treatment primarily for

   a. pathogen removal
   b. inorganic salt removal
   c. organic materials removed
   d. filtering sediments

28. To remove nitrates and phosphates from water requires

   a. aeration
   b. advanced treatment
   c. secondary treatment
   d. an activated sludge process

29. Which is not a reason that ozone is sometimes preferred over chlorine as a disinfectant for water?

   a. ozone is cheaper
   b. ozone is better for killing viruses
   c. chlorine imparts an unpleasant taste to water
   d. chlorine forms toxic by-products

30. A farmer's cattle frequently abort spontaneously. Some baby pigs are born blue in color. His well water is examined and found to be contaminated with

   a. nitrates
   b. phosphates
   c. lead
   d. bacteria

31. Acid rain isn't much of a problem in lakes in areas where there are

   a. few people
   b. few cars
   c. limestone rocks
   d. marble buildings

32. A stream is found to be highly acidic. It is probably contaminated by
   a. strip mining runoff
   b. a slaughterhouse
   c. raw sewage
   d. farm runoff

33. Chlorination of drinking water
   a. removes phosphates
   b. kills all bacteria and viruses
   c. kills some bacteria
   d. makes it taste better

ANSWERS

ADDITIONAL PROBLEMS

1. 600,000,000 kg or 600,000 t

2. 170 years (but the population and rate of use are increasing)

SELF-TEST

1. d	9. d	17. d	25. d	33. c
2. d	10. d	18. c	26. b	
3. a	11. a	19. a	27. c	
4. d	12. c	20. a	28. b	
5. c	13. b	21. d	29. a	
6. b	14. c	22. c	30. a	
7. d	15. b	23. a	31. c	
8. d	16. c	24. b	32. a	

CHAPTER 14

Energy

KEY TERMS

biomass
bitumen
breeder reactor
catalytic reforming
coal
endothermic reaction
entropy
exothermic reaction
First law of Thermodynamics

fossil fuel
fuel
fuel cell
gasoline
geothermal energy
isomerization
kerogen
Law of Conservation of Energy
natural gas

nuclear reactor
octane rating
oil shale
petroleum
photosynthesis
photovoltaic (solar) cells
plasma
Second Law of Thermodynamics
tar sands

CHAPTER SUMMARY

14.1   Heavenly Sunlight Flooding the Earth with Energy
   A. Nearly all the energy available to us on Earth comes from the sun (173,000 TW), where it is generated
      by nuclear fusion.
      1. SI unit for energy is joule (J).
      2. SI unit for power is watt (W).
      3. 1 watt = 1 joule/sec.
   B. Energy and the Life-Support System
      1. The energy the Earth receives from the sun is
         a. Reflected back into space (30%).
         b. Used to power the water cycle (23%).
   C. Green plants use sunlight (0.02%) to convert solar energy to chemical energy (photosynthesis).
      1. The sunlight is absorbed by green plant pigments called chlorophylls.
      2. The energy is used to convert carbon dioxide and water to glucose and oxygen.

14.2   Energy and Chemical Reactions
   A. The rates of chemical reactions depend on
      1. <u>Temperature</u>: The higher the temperature, the faster the reaction.
         a. At higher temperatures, molecules move faster and collide more often. Also, more energy is
            available for breaking chemical bonds.
      2. <u>Concentration of reactants</u>: The more molecules there are in a given volume of space, the more
         likely they are to collide.
         a. The more collisions there are, the more reactions are likely to occur.
      3. <u>Catalysts</u>: These substances speed up reactions without being used up in the process.
         a. Enzymes are biological catalysts that mediate the reactions in living cells.
   B. Energy changes and chemical reactions are quantitatively related to the amounts of chemicals
      undergoing change (see Example Problems).
      1. <u>Exothermic</u> reactions are those that release heat.

2. <u>Endothermic</u> reactions are those that require a net input of energy.

14.3 Energy and the First Law: Energy Is Conserved, Yet We Are Running Out
A. The First Law of Thermodynamics (also called the Law of Conservation of Energy) states that energy is neither created nor destroyed (although it can be changed in form).
B. We can't make machines that create energy from nothing, but neither does energy disappear.

14.4 Energy and the Second Law: Things Are Going To Get Worse
A. No engine can operate at 100% efficiency; some energy is converted to heat or friction.
B. Not all forms of energy are equal; high-grade forms are constantly degraded to low-grade forms.
    1. Mechanical energy (high grade) is eventually changed to heat energy (low grade).
C. The Second Law of Thermodynamics states that
    1. Energy flows spontaneously from a hot object to a cold one.
    2. Natural processes tend toward greater entropy (more disorder).
        a. We can reverse the tendency towards disorder but it costs energy to do it.

14.5 People Power: Early Uses of Energy
A. Primitive people obtained their energy (food and fuel) by hunting and gathering.
B. Domestication of animals increased available energy somewhat; people gained horsepower and oxpower.
C. Windmills and waterwheels further increased available energy.
    1. They convert energy into useful work.
D. Since 1850, steam engines and other mechanical devices have provided us with 10,000 times as much energy as was available to primitive people.

14.6 Fossil Fuels
A. Fuels are substances that burn, releasing significant amounts of energy.
    1. Fuels are <u>reduced</u> forms of matter.
    2. Burning is the <u>oxidation</u> of fuels.
B. Fossil fuels include coal, petroleum, and natural gas.
    1. Fossil fuel reserves are rapidly being depleted.

14.7 Coal: The Carbon Rock of Ages
A. Coal is a complex material; its energy content is closely related to its carbon content.
    1. Coal was formed millions of years ago from plant material buried under mud.
        a. Cellulose of plants was compressed; it broke down releasing hydrogen and oxygen and leaving behind a material rich in carbon.
    2. We use fossil fuels 50,000 times as fast as they are being formed.
    3. Of all the fossil fuels that ever existed, we will have used about 90% in 300 years.
B. Inconvenient Fuel
    1. Coal is mined and shipped to factories.
C. Abundant Fuel
    1. Coal is our most plentiful fossil fuel; the United States is estimated to have 40% of the world's reserves.
    2. Electric utilities burn 700 billion tons of coal per year to generate 55% of our electricity.
    3. Coal is our dirtiest major fuel.
        a. Most is obtained by strip mining.
D. Sources of Pollution
    1. Burning coal produces carbon dioxide, carbon monoxide, sulfur oxides (and sulfuric acid), and particulate matter.
    2. Coal can be cleaned before burning by the flotation method.

a. Coal has a density of 1.3 g/cm³. It can be floated away, leaving denser sulfur-containing minerals behind.
   E. Sources of Chemicals
      1. Coal tar (a source of chemicals) is made by heating coal to drive off volatile matter.
         a. Coal tar and liquid coal oil are good sources of organic materials.

14.8   Natural Gas: Mostly Methane
   A. Composition of natural gas
      1. Methane   60-80%
      2. Ethane   5–9%
      3. Propane   3–18%
      4. Butane and pentane   2–14%
   B. Most natural gas is burned as fuel, but some is separated into fractions.
      1. The fractions are cracked to produce ethylene, propylene, and other valuable chemical intermediates.
         a. Ethylene, propylene and four-carbon compounds are also made by cracking the alkane mixture.
         b. Natural gas is the starting material for many one-carbon compounds.
   C. Natural gas is the cleanest fossil fuel in regard to pollution.

14.9   Petroleum: Liquid Hydrocarbons
   A. Petroleum is a mixture of organic compounds.
      1. Hydrocarbons, including alkanes and cycloalkanes.
      2. Varying proportions of sulfur, nitrogen, and oxygen-containing compounds.
   B. Separation into Fractions
      1. Cracking: breaking bigger molecules into smaller ones.
         a. Gasoline
         b. Unsaturated hydrocarbons: starting materials for other needed materials.
      2. Plastics, antibiotics, pain killers, and preservatives.
   C. Combustion of hydrocarbons produces carbon dioxide, water, and heat.
      1. $2 C_8H_{18} + 25 O_2 \rightarrow 16 CO_2 + 18 H_2O$
   D. Burning petroleum leads to air pollution due to the following:
      1. Presence of other substances in petroleum, like sulfur.
      2. Inefficient burning of gasoline by internal combustion engines in cars.
   E. Petroleum is more abundant than natural gas.
   F. Gasoline
      1. Mixture of hydrocarbons from C_5H_{12} to $C_{12}H_{26}$.
         a. Straight-run gasoline: petroleum fraction straight from the distillation column.
   G. Octane Rating of Gasoline
      1. Arbitrary scale used to rate performance of gasoline.
         a. More branching in hydrocarbon chain, the higher the octane rating.
      2. Alkylation: combination of small hydrocarbon molecules into larger ones for use as fuel.
   H. Octane boosters
      1. Compounds added to gasoline to improve antiknock quality.
         a. Tetraethyllead: effective octane booster but banned due to toxicity of lead.
         b. Methyl-tert-butyl-ether: additive of choice. Contains oxygen and is referred to as <u>oxygenate.</u>

14.10   Convenient Energy: Electricity
   A. Electricity is perhaps the most useful form of energy.
      1. It flows through wires and can be converted into light, heat, or mechanical energy.
   B. Fuels are burned to make steam to turn a turbine to generate electricity.
   C. Coal-burning electric power plants are, at best, about 40% efficient; the rest of the energy is wasted as heat (thermal pollution).

14.11 Nuclear Power
  A. Energy released during fission is used to produce steam, which turns a turbine generating electricity.
    1. About 20% of United States' electricity comes from nuclear power plants.
  B. Types of nuclear plants
    1. Pressurized water reactors are common in the United States.
  C. Nuclear power plants use the same fission reactions as nuclear bombs, but these reactions are controlled.
    1. A <u>moderator</u>, such as water or graphite, is used to slow down the fission neutrons.
    2. <u>Control rods</u>, made of boron or cadmium steel, are inserted to absorb neutrons and stop the fission reaction. Partial removal of the rods starts the chain reaction.
  D. Nuclear Advantage: Minimal Air Pollution
    1. Advantages: No soot, fly ash, sulfur dioxide, or other chemical air pollutants are emitted.
      a. Nuclear power contributes nothing to global warming, air pollution, and acid rain.
  E. Problems with Nuclear Power
    1. Elaborate safety precautions are required.
    2. Runaway nuclear reactions are unlikely, but possible.
    3. Nuclear wastes are highly radioactive and must be isolated for centuries.
      a. Tailings from uranium mines are mildly radioactive and contaminate wide areas.
      b. Slightly more thermal pollution is generated from nuclear power plants than that from fossil fuel-burning plants.
  F. Nuclear Accidents: Real and Imagined Risks
    1. Nuclear power plants cannot blow up like nuclear bombs. The uranium-235 is enriched, at most, to 3-4% for power generation. Nuclear bombs require enrichment to about 90%.
  G. Breeder Reactors: Making More Fuel Than We Burn
    1. Less than 1% of natural uranium is the fissile uranium-235 isotope; the rest is non-fissile uranium-238.
    2. Breeder reactors have a core of fissile material surrounded by uranium-238. Neutrons from the core convert the uranium-238 to fissile plutonium-239. The process breeds more fuel than it consumes.
      a. Advantage: There is enough uranium-238 to last for centuries.
      b. Disadvantages: Plutonium melts at 640°C; operation is limited to rather cool and inefficient temperatures.
        i. Molten sodium metal, which reacts violently with both air and water, is used as a coolant.
        ii. Core meltdown is more likely than with uranium-235 fuel.
        iii. Plutonium is toxic.
        iv. Reactor-grade plutonium can be converted to nuclear bombs.
    3. Breeder reactors can also convert non-fissile thorium-232 to fissile uranium-233.
      a. Uranium-233 emits biologically damaging alpha particles.
    4. No breeder reactors are operating in the United States, but they are used in France and other countries.

14.12 Nuclear Fusion: The Sun Is a Magnetic Bottle
  A. Thermonuclear reactions power the sun and hydrogen bombs, but controlled, sustainable fusion reactions are yet to be achieved.
  B. Controlled fusion would have several advantages.
    1. The principal fuel, deuterium (^{2}H) is plentiful.
    2. The product, helium, is biologically inert; radioactive wastes are minimized.
  C. Some possible disadvantages
    1. Radioactive tritium (^{3}H) might be released and incorporated into living organisms.
    2. Temperatures of 50 million °C are required.
    3. The plasma, which is a mixture of electrons and nuclei, must be contained by a magnetic field or other nonmaterial device.
  D. It is unlikely that electricity from fusion will be available until well into the next century.

14.13 Harnessing the Sun: Solar Energy
   A. Energy from the sun is diffuse and must be concentrated to be useful.
   B. Solar Heating: Solar collectors can be used to heat homes and water for bathing, laundry, etc.
   C. Solar Cells: Electricity from Sunlight
       1. Photovoltaic cells (solar cells) can convert sunlight directly to electricity.
       2. Solar cells based on silicon have
           a. Donor crystals doped with arsenic to provide extra electrons (silicon has four valence electrons, arsenic five).
           b. Acceptor crystals doped with boron (three valence electrons) create positive holes.
           c. Sunlight dislodges electrons from donor crystals, creating a current flow from donor cell to receptor crystals.
   D. Solar cells have low efficiency (about 10%).
   E. Solar energy is not available at night and on cloudy days, but it can be stored, for example, as heat, in molten salts.

14.14 Biomass: Photosynthesis for Fuel
   A. Biomass (plants grown for fuel) has several advantages.
       1. It is a renewable resource.
       2. Energy for biomass production comes from the sun.
   B. Disadvantages
       1. Most of the available land is needed for food production.
       2. Plants must be harvested and hauled to where the energy is needed, often over long distances.
   C. Biomass can be burned directly as fuel or converted to other fuels.
       1. Plants high in starches and sugars can be used to produce ethanol.
       2. Wood can be used to produce methanol.
       3. Bacteria can convert plant material to methane.

14.15 Other Energy Sources
   A. Wind Power and Water Power
       1. The sun causes the wind to blow and water to evaporate and fall later as rain.
           a. Blowing wind and flowing water can be used as energy sources.
       2. About 10% of our electricity comes from hydroelectric plants, but nearly all of the good dam sites are in use.
       3. Windmills can be used to provide mechanical energy or to generate electricity.
           a. Windmills could meet up to 10% of our electricity needs, but provide an almost negligible amount at present.
       4. Wind and water provide relatively clean energy, but suffer from some disadvantages.
           a. Dams flood valuable land. Reservoirs silt up. Sometimes dams break, causing devastating floods.
           b. The wind doesn't always blow; storage or an alternate source is needed. Windmills limit the use of the land on which they are placed.
   B. The Tides: Moon Power
       1. Tides possess great energy; one tidal-power plant is in operation in France.
       2. Few sites are suitable and those that are, are known for their scenic beauty.
           a. Energy generation is possible only when the tides are going out or coming in.
           b. Storage devices or alternate sources are needed.
   C. Geothermal Energy
       1. The interior of the Earth is hot; in some areas the hot parts come to the surface as geysers or volcanoes.
       2. Disadvantages: Waste water is quite salty.
   D. Oil Shale and Tar Sand
       1. The organic material in oil shale is kerogen.

      a. When heated, kerogen breaks down, forming hydrocarbons similar to those in petroleum.
    2. Tar sands contain bitumen, a mixture of heavy hydrocarbons.
    3. The "oil" in oil shale and tar sands is high-entropy material. It takes a lot of energy to convert it to useful, low-entropy forms.
  E. Oil from Seeds
    1. Fast-growing plants with high oil yields may be developed for producing fuel in the future.
  F. Coal Gasification and Liquefaction
    1. Coal can be converted to a fuel similar to natural gas. The essential process is reduction of carbon to hydrogen to form methane.

$$C + 2H_2 \rightarrow CH_4$$

    2. Coal can be converted to liquids. In one process, coal is first converted to methanol, and then the methanol is converted to gasoline.
    3. Coal liquefaction and gasification have several disadvantages.
      a. A significant portion of the energy is wasted in the conversion process.
      b. Liquid fuels from coal contain sulfur, nitrogen and arsenic compounds, which contribute to air pollution.
    4. The Bergius and Fischer-Tropsh methods are also used to make liquid fuel from coal.
  G. Hydrogen as Fuel
    1. On weight basis, hydrogen has more energy than most other fuels.
  H. Alcohol as Fuel
    1. Gasohol: ethanol mixed with gasoline.
  I. Fuel Cells
    1. Fuel cell fuel is oxidized in an electrochemical cell.
      a. They differ from electrochemical cells in two ways.
        i. Fuel and oxygen are fed in continuously.
        ii. The electrodes are an inert material such as platinum.
      b. In a hydrogen fuel cell, hydrogen is oxidized at the anode, and oxygen is reduced at the cathode.
      c. Fuel cells are used on spacecraft, but they contribute little to the production of electricity on Earth.

14.16 Energy: How Much Is Too Much?
  A. Energy production cannot be pollution free.
  B. Thermal pollution has already modified the climate around major metropolitan areas.
  C. We can minimize the negative effects of energy production by conserving energy and by making the wisest possible choices of production methods.
    1. Use of energy-efficient appliances
    2. Use of public transportation

CHAPTER OBJECTIVES

(You should...)

1.   Be able to define or identify each of the key terms.

2.   Be able to describe the Earth's incoming energy vs. outgoing energy (Table 14.1 of the text).

3.   Know that all our significant energy sources except nuclear and geothermal energy come ultimately from the sun.

4.   Know that energy for maintenance of life on Earth comes almost entirely from photosynthesis—either that taking place today or that which occurred ages ago.

5.	Know the First Law of Thermodynamics: Energy is conserved.

6.	Know the Second Law of Thermodynamics:
	Energy flows from hot to cold
	Systems go from low to high entropy

7.	Understand the effect of temperature, concentrations of reactants and catalysts on chemical reactions.

8.	Be able to discuss the history of humanity's use of energy, sources of energy, where used, advantages and disadvantages, and amounts of energy used.

9.	For each type of fossil fuel (gas, oil, and coal) be able to
	a) identify the principle chemical
	b) identify the advantages and disadvantages
	c) write balanced chemical equations for their combustion
	d) identify the percentage of energy each produces
	e) identify how or where each fossil fuel is used

10.	Know that coal is the most abundant fuel in the United States (and the world), not very convenient to use, composed of mostly carbon, a major source of air pollution unless "cleaned," and a good source of organic compounds.

11.	Know that natural gas is composed mostly of methane, a source for many one-carbon compounds, and the cleanest burning fuel.

12.	Know that petroleum is a mixture of liquid hydrocarbons, separated into fractions for gasoline, kerosene, fuel oil etc., can be modified in "cracking" towers, and is a major source of unsaturated hydrocarbons for polymers.

13.	Know that electricity is a convenient secondary source of energy but that the conversion to electricity from fossil fuels is inefficient and polluting.

14.	Know that 20% of the energy in the United States comes from nuclear energy, the basics of the operation of a nuclear power plant, and be able to discuss the advantages and disadvantages of nuclear fission power plants.

15.	Be able to discuss the basics of solar heating and photovoltaic cells and discuss their advantages and disadvantages.

16.	Be able to discuss the basics of biomass energy and discuss their advantages and disadvantages.

17.	Be able to discuss the basics of other sources of energy: wind and water power; tides; geothermal energy; oil shale and tar sands; oil from seeds, coal gasification and liquefaction; hydrogen; alcohol; and fuel cells. Be able to discuss their advantages and disadvantages.

18.	Know that there is a limit to how much energy Earth can use.

EXAMPLE PROBLEMS

1.	Complete combustion of 16.0 g of methane yields 192 kcal of energy. How much energy is obtained by the combustion of 96.0 g of methane?

$$96.0 \text{ g methane} = 192 \text{ kcal}$$

$$16.0 \text{ g methane} = 1150 \text{ kcal}$$

2.   To split 36.0 g of water into hydrogen and oxygen requires the input of 137 kcal of energy. How much energy is required to split 180 g of water?

$$180 \text{ g water} = 137 \text{ kcal}$$

$$36.0 \text{ g water} = 685 \text{ kcal}$$

ADDITIONAL PROBLEMS

1.   Burning 4.00 g of hydrogen in sufficient oxygen produces 137 kcal of heat. How much heat is released by the combustion of 20.0 g of hydrogen?

2.   How much carbon dioxide is formed by the complete combustion of 78 g of carbon?

$$\text{The equation is } C + O_2 \rightarrow CO_2.$$

3.   Complete combustion of 16.0 g of methane yields 192 kcal of energy. How many grams of methane are needed to produce 1850 kcal of energy?

4.   A lot of coal contains 3.0% sulfur. How much sulfur is there in a metric ton (1000 kg) of coal? How much sulfur dioxide is formed by burning this coal? The equation is $S + O_2 \rightarrow SO_2$.

5.   In 1975 nuclear power accounted for about 0.02×10^{12} watts of electricity and the formation of tritium, with an activity of about 7.5×10^{15} disintegrations per second. By the year 2000, it is estimated that nuclear power will produce 1×10^{12} watts of electricity per year. What activity of tritium will be produced?

6.   Nuclear power plants producing 1×10^{12} watts of electricity will produce 9000 kg of waste strontium-90. The half-life of this isotope is 28 years. If 9,000 kg are produced in the year 2000, how much of that strontium-90 will remain in 2028? In 2056? In 2112?

7.   A large power plant will burn 25 million kg of coal per day. If the coal is 8% inorganic ash, how much ash is produced per day? Per year? Per decade?

8.   If the coal in problem 7 contains 0.2 part per million of mercury (that is, 0.2 kg of mercury in 1,000,000 kg of coal), how much mercury is released into the environment each day? Each year?

DISCUSSION

Energy, the ability to do work, ultimately comes from the sun. All of the food for life comes from photosynthesis that is powered by the sun. Rates of chemical reactions are increased with increasing temperature, increasing concentration, and the presence of a catalyst. Almost all chemical reactions use energy (endothermic) or give off energy (exothermic). The First Law of Thermodynamics states that energy is conserved. The Second Law of Thermodynamics states that energy flows from high energy to low energy and that disorder increases. Since the 1850s, humanity has had a tremendous increase in the amount of energy available so that now we have 10,000 times as much energy as our ancestors. Ninety percent of our energy comes from fossil fuels. The fossil fuels are coal, petroleum, and natural gas and they are comprised of organic compounds that are burned to produce energy. Each has its advantages and disadvantages. Because of its convenience, a lot of energy is converted to electricity, a secondary energy source. While nuclear power is cleaner it has the disadvantages of radiation. There are a number of

168

"alternative" energy sources, each with its advantages and disadvantages: solar energy, biomass energy, wind and water power, geothermal energy, oil shale, and tar sands. Oil from seeds, coal, gasification and liquefaction, hydrogen, alcohols and fuel cells. Since any production of energy has its disadvantages, wise stewardship of energy is necessary.

SELF-TEST

Multiple Choice

1. Energy for maintenance of life on Earth comes from

   a. nuclear reactions in the Earth's interior
   b. the hydrologic cycle
   c. photosynthesis
   d. fossil fuels

2. Prior to the 1800s, the principal fuel being used was

   a. coal
   c. natural gas
   b. oil
   d. wood

3. The First Law of Thermodynamics states that

   a. matter is conserved
   c. matter can be changed into energy
   b. energy is conserved
   d. entropy is conserved

4. The Second Law of Thermodynamics states that

   a. isolated systems do not proceed spontaneously from disorder to order
   b. all the chemical energy in oil can be converted into electrical energy
   c. the universe is proceeding toward a more ordered state
   d. the temperature of the Earth is increasing

5. Which of the following sources of power will likely last the longest?

   a. coal
   c. uranium-235
   b. petroleum
   d. the sun

6. To change a system from a disordered state to a more ordered state involves the input of

    a. energy   b. matter
    c. light   d. heat

7. Which of the following systems has the most entropy?

    a. OOOOO OXOXO XXXXX XOXOX   b. OOOOO XXXXX OOOOO XXXXX
    c. OXOOX XOXXO XXOXO OOXOX   d. OOOXX OOOXX OOXXX OOXXX

8. Coal is our most plentiful fossil fuel. The most serious disadvantage to burning coal is that it

    a. doesn't produce much heat
    b. produces a lot of sulfur oxides
    c. produces a lot of peroxyacetyl nitrate (PAN)
    d. produces a lot of ozone

9. Which physical state has the greatest entropy?

    a. gas   b. liquid
    c. solid   d. all have the same entropy

10. Which of the following is not powered by the sun?

    a. the water cycle   b. the winds
    c. photosynthesis   d. geothermal energy

11. The fossil fuel whose reserves are currently thought to be in the shortest supply is

    a. natural gas   b. nuclear power
    c. coal   d. petroleum

12. Combustion of a hydrocarbon involves its reaction with

    a. acetylene   b. an acid
    c. hydrogen   d. oxygen

13. High-grade coal is mainly

    a. methane   b. a mixture of hydrocarbons
    c. carbon   d. sulfur

14. Gasoline is a

    a. single, complex hydrocarbon
    b. mixture of a few simple hydrocarbons
    c. mixture of a few complex hydrocarbons
    d. mixture of a large number of hydrocarbons

15. Which of the following compounds has the highest octane number?

a. $CH_3CH_2CH_2CH_2CH_2CH_3$

b. 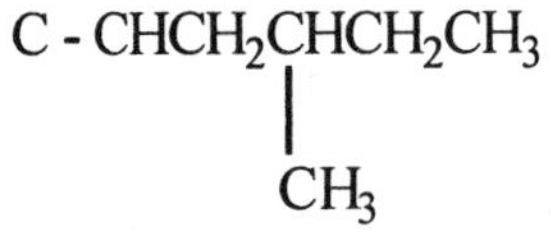 C 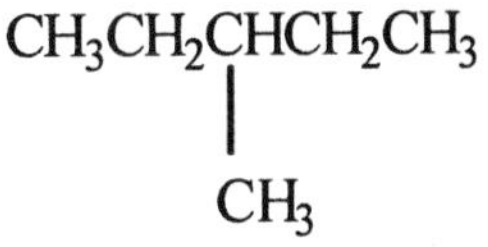 - CHCH$_2$CHCH$_2$CH$_3$
$\qquad\qquad$ |
$\qquad\qquad$ CH$_3$

c. CH$_3$CH$_2$CHCH$_2$CH$_3$
$\qquad\qquad$ |
$\qquad\qquad$ CH$_3$

d. 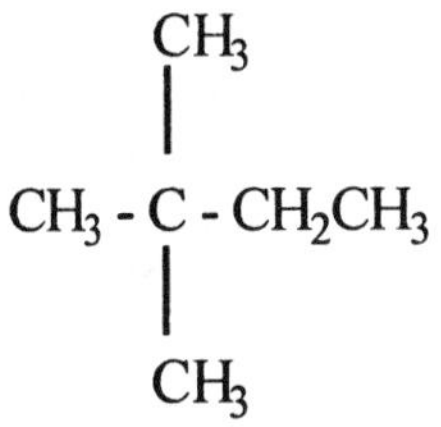 $\qquad\quad$ CH$_3$
$\qquad\qquad$ |
$\quad$ CH$_3$ - C - CH$_2$CH$_3$
$\qquad\qquad$ |
$\qquad\qquad$ CH$_3$

16. Which of the following is not a fuel?

a. C
b. CH_4
c. H_2
d. CO_2

17. The SI unit of energy is the

a. calorie
b. kilocalorie
c. joule
d. watt

18. A reaction that results in the release of heat is said to be

a. catalytic
b. endothermic
c. exothermic
d. impossible

19. Which of the following is not a fossil fuel?

a. coal
b. hydrogen
c. natural gas
d. petroleum

20. No matter what the energy source is, all power plants have problems with thermal pollution because

a. all energy transformations require matter
b. fossil fuels cause air pollution
c. all forms of energy can be transformed into heat
d. the required energy transformations are inefficient

21. Which form of energy is most convenient?

a. coal
b. electricity
c. heat
d. petroleum

22. What proportion of electricity in the United States is generated by burning coal?

a. 3%
b. 5%
c. 35%
d. 56%

23. The maximum efficiency of a coal-burning power plant is about

      a. 10%   b. 20%
      c. 40%   d. 80%

24. Which of the following does not represent a problem associated with nuclear fission reactors used as energy sources?

      a. limited supply of fuel
      b. likelihood of an atomic-bomb-like explosion
      c. disposal of exhausted fuel
      d. disposal of tailings from uranium mines

25 What proportion of electricity in the United States is generated by nuclear power plants?

      a. 1%   b. 7%
      c. 20%   d. 71%

26. Nuclear bombs can be made from reactor-grade

      a. uranium-235   b. uranium-238
      c. plutonium-239   d. all of these

27. A breeder reactor creates more fuel than it burns by changing

      a. uranium-238 into uranium-235
      b. uranium-238 into plutonium-239
      c. thorium-234 into uranium-235
      d. energy into matter

28. A fusion reactor is not presently used as an energy source because

      a. a source of fuel has not been located yet.
      b. the chemistry of the fusion process has not been worked out.
      c. the required temperature for initiating the fusion reaction is not known.
      d. a procedure for achieving the necessary conditions in a controlled manner has not been worked out yet.

29. Which type of reactor now contributes significantly to the production of electrical energy in the United States?

      a. breeder   b. fission
      c. fusion   d. all of these

30. Which process powers the sun?

      a. alpha decay   b. combustion of hydrogen
      c. fission   d. fusion

31. Controlled fusion promises plentiful power

    a. by 2000
    b. by 2005
    c. by 2010
    d. sometime in the mid-twenty-first century

32. Tailings from uranium mines give off

    a. cesium-137   b. radon gas
    c. sulfur oxides   d. tritium

33. Which solar technology is available today for widespread use?

    a. heat from solar collectors
    b. electricity from orbiting stations
    c. sun-powered automobiles
    d. electricity from production of hydrogen

34. Wind power, if fully utilized, could supply approximately what percent of our present energy needs?

    a. 1   b. 10
    c. 50   d. 90

35. How efficient is the direct conversion of sunlight to electricity at present?

    a. 1%   b. 10%
    c. 44%   d. 90%

36. Hydroelectric power provides about 10% of our energy needs at present. What percent is it expected to provide by the year 2050?

    a. 4   b. 10
    c. 20   d. 50

37. A big disadvantage of geothermal energy is that

    a. it is radioactive
    b. it releases sulfur dioxide
    c. large quantities of salty water must be disposed of
    d. it releases tritium

38. The main opposition to harnessing the tides for energy is that

    a. the moon rises at different times during the year
    b. large quantities of salty water must be disposed of
    c. scenic beauty is marred
    d. the ocean is filled with silt

39. The use of biomass for fuel is limited mainly by the availability of

    a. sunlight   b. water
    c. land   d. power plants to burn it

40. The use of oil shale is limited by the fact that

    a. its extraction involves strip mining
    b. its production requires a lot of water
    c. there is a low return for the energy invested
    d. all of these

41. The gasification of coal usually involves the reduction of carbon to

    a. methane   b. carbon dioxide
    c. sulfur dioxide   d. propane

42. Hydrogen gas is a desirable fuel because

    a. it is quite inexpensive
    b. it is readily extracted from volcanic gases
    c. the product of its combustion is water
    d. it can be obtained from water with little expenditure of energy

43. Solar cells are now widely used to provide energy for

    a. nations   b. cities
    c. homes   d. electronic calculators

44. Which of the following is a renewable energy source?

    a. biomass   b. coal
    c. natural gas   d. tar sands

45. A device that produces electricity directly from fuel and oxygen is called a

    a. combustion chamber   b. fuel cell
    c. solar cell   d. turbine

ANSWERS

Additional Problems

1. 685 kcal
2. 290 g CO_2
3. 154 g methane
4. 30 kg S; 60 kg SO_2
5. 3.75×10^{17} disintegrations/sec
6. 4500 kg; 2250 kg; 562.5 kg
7. 2 million kg/day; 730 million kg/year; 7300 million kg/decade
8. 5 kg/day; 1825 kg/year

Self-Test

1. c	9. a	17. c	25. c	33. a	41. a
2. d	10. d	18. c	26. c	34. b	42. c
3. b	11. a	19. b	27. b	35. b	43. d
4. a	12. d	20. d	28. d	36. a	44. a
5. d	13. c	21. b	29. b	37. c	45. b
6. a	14. d	22. d	30. d	38. c	
7. c	15. d	23. c	31. d	39. c	
8. b	16. d	24. b	32. b	40. c	

CHAPTER

15

Biochemistry

A Molecular View of Life

KEY TERMS

active site	coenzyme	messenger RNA (mRNA)	primary structure
alpha helix	cofactor	metabolism	protein
amino acid	disaccharides	monosaccharide	purine
anabolism	disulfide linkage	nucleic acid	pyrimidine
anticodon	enzyme	nucleotides	replication
apoenzyme	fat	oils	secondary structure
base triplet	fatty acid	peptide bond	tertiary structure
biochemistry	gene	pleated sheet	transcription
carbohydrate	globular protein	polymerase chain reaction (PCR)	transfer RNA (tRNA
catabolism	iodine number	polypeptide	translation
condon	lipid	polysaccharide	zwitterion

CHAPTER SUMMARY

15.1   The Cell: The Biological Unit of Life
  A. Cells are enclosed in membranes.
      1.   Plant cells also have cell walls.
  B. Cell structure
      1.   Cell membrane: The cell gains nutrients and gets rid of wastes through the cell membrane.
      2.   Cell nucleus contains nucleic acids that control heredity.
      3.   Ribosomes are the location of protein synthesis.
      4.   Chloroplasts (plant cells)—energy is absorbed and converted to chemical energy.

15.2.   Energy in Biological Systems
  A. Living organisms can use only certain forms of energy.
  B. Plants' chloroplasts convert radiant energy into chemical energy stored as carbohydrates (glucose).
  C. Animals cannot use sunlight directly. Animals obtain energy from
      1.   Carbohydrates.
      2.   Fats.
      3.   Proteins.
  D. Metabolism—chemical reactions that keep cells alive
      1.   Catabolism—degradation of molecules to provide energy.
      2.   Anabolism—synthesis of biomolecules.

15.3 Carbohydrates: A Storehouse of Energy
   A. Carbohydrates are compounds of carbon, hydrogen, and oxygen.
   B. Some simple sugars
      1. Monosaccharides cannot be hydrolyzed (split apart by water).
         a. Some examples: glucose (also called dextrose), fructose (fruit sugar), and galactose (milk sugar).
      2. Disaccharides can be hydrolyzed into two monosaccharides.
         a. Some examples
            i. Sucrose → glucose + fructose
            ii. Lactose → glucose + galactose
   C. Polysaccharides: Starch and Cellulose
      1. Polysaccharides yield many monosaccharide units upon hydrolysis.
         a. Some examples: starch and cellulose.
            i. Both starch and cellulose are polymers of glucose.
      2. In starch, glucose molecules are connected through an alpha linkage where the oxygen atom joining the glucose molecules is pointing downward.
         a. Two kinds of plant starch are
            i. Amylose where glucose units are joined in a continuous chain.
            ii. Amylopectin where glucose units are joined in a branched chain.
         b. Animal starch is glycogen.
            i. In glycogen glucose units are joined in a branched chain.
      3. In cellulose, glucose molecules are connected through a beta linkage where the oxygen atom joining the glucose molecules is pointing upward.

15.4 Fats and Other Lipids
   A. Lipids are defined by their solubilities, not their structure.
      1. Lipids are not soluble in water but are soluble in organic solvents.
      2. Some examples of lipids: fats, fatty acids, steroids. sex hormones, and fat soluble vitamins.
   B. Fats are esters of fatty acids and glycerol (a trihydroxy alcohol).
      1. Fats are classified according to the number of fatty acid chains.
         a. Naturally occurring fatty acids have an even number of carbon atoms.
      2. Animal fats are usually solid at room temperature and have a higher proportion of saturated fatty acids.
         a. Saturated fatty acids do not have carbon-carbon double bonds.
         b. Saturated fats have a high proportion of saturated fatty acids.
      3. Vegetable fats (oils) are liquid at room temperature and have a higher proportion of unsaturated fatty acids.
         a. Polyunsaturated fatty acids have two or more carbon-carbon double bonds.
         b. Polyunsaturated fats have mostly polyunsaturated fatty acids.
      4. The degree of unsaturation of a fat is measured by the iodine number.
         a. Iodine number is the number of grams of iodine that will be added to the double bonds by 100 g of fat.
         b. The higher the iodine number, the more unsaturated the fat.
      5. Fats and oils are insoluble in water and because they are lighter, they float.

15.5 Proteins: Polymers of Amino Acids
   A. No living part of humans or any other organism is without protein.
      1. Proteins serve as the structural material of animals.
         a. Proteins are made of carbon, hydrogen, oxygen, nitrogen and often sulfur.
   B. Proteins are copolymers of 20 different amino acids (see Table 15.3 in the text).

1. Amino acids have two functional groups, an amino group (—NH₂) located on the carbon next to the carbon of the carboxyl group (—COOH).

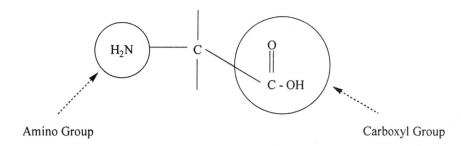

Amino Group   Carboxyl Group

   a. The carboxyl group is acidic and donates a H⁺ to the amino group, which is basic and accepts the H⁺. This forms a <u>zwitterion</u>.

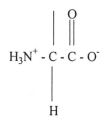

   b. A zwitterion is a compound in which the anion and cation are part of the same molecule.
   c. Amino acids differ in the variety of other groups attached to the central C.
C. Plants synthesize proteins from CO_2, water, and minerals (supplying N and S). Animals must take in proteins as food.

15.6   The Peptide Bond: Peptides and Proteins
   A. Proteins are polyamides: amino acids linked by many peptide bonds (amide linkages).
      1. A peptide bond (amide linkage) links the —NH₃⁺ of one amino acid with the —COO⁻ of another amino acid.

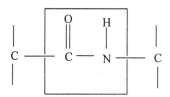

Peptide Bond

      a. This arrangement leaves a free carboxyl group (—COO⁻) at one end of the protein called the C-terminal and a free amino group (—NH₃⁺) at the other end called the N-terminal .
      2. Dipeptides are links of two amino acids, tripeptides are links of three, and polypeptides are links of ten or more amino acids.
      a. A protein is a polypeptide that has a molecular weight of more than 10,000 amino acid units.
   B. Sequence of amino acids
      1. The sequence in which the amino acids are connected in peptides is of critical importance.
         a. Sequences are written using three-letter abbreviations for the amino acids and arranging them with the N-terminal to the left of the sequence and the C-terminal to the right.
         b. A minor change in the amino acid sequence of a protein may have disastrous effects for the organism.

1. Example: Sickle cell anemia is caused by one incorrect amino acid in a 300-unit sequence.

15.7 Structure of Proteins
A. Primary structure of a protein is its amino acid sequence.
B. Secondary structure is the arrangement of chains about an axis.
  1. Pleated-sheet structure: Molecules are stacked in extended arrays with hydrogen bonds holding adjacent chains together. (An example is silk.)
  2. Alpha-helix structure: A right-handed helix formed when the amino group ($-NH_3^+$) of one amino acid in one turn of the chain forms hydrogen bonds with the carboxyl ($-COO^-$) group of another amino acid. This arrangement allows the protein to be stretched and then regain its shape, like a spring. (An example is wool.)
C. Tertiary structure describes how the protein chain is folded.
  1. Example: globular proteins
D. Quaternary structure describes the packing of proteins that have more than one polypeptide chain.
E. Four Ways to Link Protein Chains
  1. Peptide bonds fix the primary protein structure.
  2. Four other forces hold the protein in a structural arrangement.
    a. Hydrogen bonding—the carbonyl oxygen of one peptide may form a hydrogen bond to an amide hydrogen (N—H).
      1. Examples: alpha helix structure of wool protein and pleated sheet structure of silk protein
    b. Salt bridge—a proton transfer between the acidic side chain of one amino acid and the basic side chain of another results in opposite charges, which then attract each other.
    c. Disulfide linkage—formed when two cysteine units (–SH) are oxidized. The resulting disulfide bond between the two cysteine units is a covalent bond that is much stronger than a hydrogen bond.
    d. Hydrophobic Interaction—weak attractive forces between nonpolar side chains.
      i. Significant in the absence of other forces.
      ii. Nonpolar side chains cluster together on the inside folds of proteins forming several hydrophobic interactions in a given region of the protein.

15.8 Enzymes: Exquisite Precision Machines
A. Enzymes are specialized proteins that act as highly specific biological catalysts.
  1. Enzymes enable reactions to occur at convenient rates and lower temperatures by changing the reaction path.
  2. The enzyme reacts with a compound (substrate) and then separates into a new product and the regenerated enzyme.
    a. The substrate must fit a portion of the enzyme (active site) precisely in order for the reaction to occur.
    b. The substrate and enzyme are held together by bonds between complementary charged groups on the two.
    c. The formation of new bonds weakens the old bonds to the substrate and facilitates the breaking of old bonds and the formation of new products.
  3. An enzyme can be made ineffective when inhibitor molecules bond to the enzyme at positions remote from the active site, thus changing the shape of the enzyme and preventing further bonding with the substrate.
  4. Inorganic ions serve as cofactors necessary for proper functioning of some enzymes.
    a. Coenzymes are organic cofactors that are not proteins. Many are derived from vitamins.
    b. The protein part of an enzyme is called the apoenzyme.

15.9   Nucleic Acids: The Chemistry of Heredity
    A. Two kinds of nucleic acid are deoxyribonucleic acid (DNA) found in the cell nucleus and ribonucleic acid (RNA) found in all parts of the cell.
        1. Nucleic acids are chains of repeating nucleotides.
            a. Nucleotides contain a sugar, a heterocyclic amine base, and a phosphate unit.
        2. The sugar is either ribose (found in RNA) or deoxyribose (found in DNA).
            a. The two sugars differ in the presence of an oxygen on the second carbon. Ribose has an oxygen in this position. Deoxyribose does not.
        3. The bases are either purines (two fused rings) including adenine and guanine or pyrimidines (one ring) including cytosine, thymine, and uracil.
        4. The phosphate groups are attached to the fifth carbon of the sugar.
            a. AMP (adenosine monophosphate) has one phosphate group.
        5. The phosphate groups of nucleotides form ester linkages to the hydroxyl groups of sugars of adjoining nucleotides to form nucleic acid chains.
        6. The chain is then composed of a phosphate-sugar backbone with branching heterocyclic bases.
        7. DNA: Sugar is deoxyribose and the bases are adenine, guanine, cytosine, and thymine.
        8. RNA: Sugar is ribose and the bases are adenine, guanine, cystosine, and uracil.
    B. Base sequence in nucleic acids
        1. The sequence of bases in huge nucleic acid molecules is used to store all the information needed to build living organisms.
    C. The Double Helix
        1. The bases in DNA are paired (adenine to thymine) and (guanine to cytosine) through hydrogen bonding.
            a. Watson and Crick determined that DNA was composed of two helixes wound around each other and held in place by base pairing.
        2. In base pairing a pyrimidine base is paired with a purine base.
            a. In the pyrimidine-purine pair of guanine and cytosine, three hydrogen bonds can form. No other such pairing will provide such extensive interaction.
    D. Structure of RNA.
        1. RNA is a single strand of nucleic acid with some internal base pairing where the molecule folds back on itself.

15.10   DNA: Self-replication
    A. Chromosomes and genes.
        1. Chromosomes contain the hereditary material.
            a. The number of chromosomes varies with species. Human cells have 46 chromosomes with egg and sperm cells each providing half these chromosomes.
            b. The nucleic acid in chromosomes is DNA.
        2. Genes are sections of the DNA molecule.
        3. During cell division, each chromosome replicates (produces an exact duplicate of) itself.
    B. Replication of DNA.
        1. Two chains of double helix are pulled apart and each chain directs synthesis of a new DNA chain.
            a. Synthesis begins with a nucleotide from the surrounding cellular fluid pairing with its complementary base on the DNA strand.
        2. As nucleotides align, the enzymes connect them to form the sugar-phosphate backbone of the new chain.
        3. When the cell divides, each daughter cell receives one of the original DNA strand.
        4. The sequence of bases along the DNA strand encodes all the information for building a new organism.
            a. These four bases in the genetic code can form many different combinations.

15.11 RNA: Protein Synthesis and the Genetic Code
  A. RNA carries the information from DNA to other parts of the cell.
      1. Transcription: Transfer of DNA information to mRNA (messenger RNA) through base pairing.
          a. In RNA, uracil takes the place of thymine so the allowed base pairing between DNA and mRNA is T-A, C-G, A-U.
      2. Translation of the code in mRNA into a protein structure takes place in the ribosomes. The mRNA becomes attached to a ribosome and the genetic code is deciphered.
  B. Transfer RNA (tRNA), located in the cytoplasm, translates the base sequence of mRNA into the amino acid sequence of a protein.
      1. tRNA has a base triplet that determines which amino acid will attach to the end of tRNA. This amino acid can then be moved into position when the tRNA base pairs with the appropriate bases on mRNA.
      2. When amino acids have been moved into position by tRNAs these amino acids then bond to each other forming a peptide chain. In this way, a protein is built up.
      3. The base triplet on the mRNA is called a codon and pairs with its complementary base triplet on tRNA called an anticodon.
      4. Some amino acids are specified by several different codons, others are specified by only one codon.
          a. Three codons are stop signals, calling for the termination of a proton chain.

15.12 Genetic Engineering
  A. RFLP: predicting genetic disease.
      1. Treating DNA with restriction endonucleases yields segments of genetic material called restriction fragment length polymorphisms (RFLPs).
          a. RFLPs contain fewer genes and demonstrate patterns that may be linked to a hereditary genetic disease.
      2. Genetic engineering may lead to the substitution of a functioning gene for a defective gene in a person's cell.
  B. Recombinant DNA
      1. A gene from one organism can be substituted for a defective or missing gene in another organism.
          a. The gene is identified, isolated using the RFLP process, and placed in a separate piece of DNA.
          b. The recombined DNA is then transferred into a bacteria DNA (plasmid). The bacteria are cloned and large amounts of protein coded by the genes are produced.
              i. Human growth hormone, epidermal growth factor, and insulin are produced using recombinant DNA.

CHAPTER OBJECTIVES

(You should be able to...)

1. Define all the key terms.

2. Identify the cell's structural components.

3. Explain the difference between catabolism and anabolism.

4. Differentiate between mono- and disaccharides.

5. Identify a simple sugar as a mono- or disaccharide.

6. Explain the difference between the polysaccharides starch and cellulose.

7. Distinguish between the two types of plant starch and one type of animal starch.

8. Define lipids by their behavior and give examples.

9. Define fats and differentiate between saturated and unsaturated fats.

10. Explain the relationship between iodine number and the degree of unsaturation of a fat.

11. Define proteins.

12. Recognize the amino and carboxyl groups on an amino acid.

13. Explain what a zwitterion is.

14. Describe how a peptide bond forms.

15. Recognize the C-terminal and N-terminal of a protein.

16. Explain the difference between a di-, tri-, and polypeptide.

17. Explain the relationship between the amino acid sequence of proteins and genetic diseases.

18. Define and differentiate between the four levels of protein structure.

19. Identify and explain the four forces that hold proteins in structural arrangements.

20. Explain what an enzyme is chemically.

21. Describe how an enzyme works.

22. Differentiate between cofactor, coenzyme, and apoenzyme.

23. Be able to name the two kinds of nucleic acids and state which kind occurs in the cell nucleus.

24. Identify the sugar unit in each kind of nucleic acid and identify each sugar from its structure.

25. List the heterocyclic bases that are found in each kind of nucleic acid.

26. Recognize a nucleotide from its structure.

27. Identify the bases that occur in nucleic acids as purines or pyrimidines.

28. Explain how bases are paired through hydrogen bonds.

29. Given a base in DNA or RNA, choose the appropriate pairing base.

30. Describe (briefly) each of the following:
    a. DNA self-replication
    b. transcription
    c. translation
    d. protein synthesis
    e. RFLPs
    f. recombinant DNA

DISCUSSION

The cell is the structural unit of life. Biochemistry, the chemistry of life, is organized into classes of compounds. Carbohydrates (sugars and starches) are the products of photosynthesis and are the source of energy for living things. Fats and lipids are water insoluble compounds with important members such as steroids (cholesterol and sex hormones). Proteins, polymers of amino acids, are the structural unit of animals. There are about 20 amino acids found in life on earth. They are bonded with the peptide link to form polymers. They also serve as enzymes or biochemical catalysts. There are three structures for each protein: primary (the sequence of amino acids), secondary (the arrangement about an axis), and tertiary (folding). Sometimes there is a quaternary structure (details the relationship between the first structures). Nucleic acids (RNA, DNA) determine the heredity of life. RNA and DNA consist of repeating units of nucleotides that are made up of a sugar, a heterocyclic base, and a phosphate. The sugar is either ribose (RNA) or deoxyribose (DNA). The sequence of bases in the chain determines the genetic code. The two strands of DNA (RNA) are held together with H-bonding. These are broken and new ones formed during replication (reproduction). Recent advances in identifying DNA (DNA fingerprinting) and modifying the DNA (recombinant DNA) are modern issues.

SELF-TEST

Multiple Choice

1. Green plants make food by using their

    a. ribosomes
    c. nucleus
    b. cell walls
    d. chloroplasts

2. Plants store chemical energy as

    a. glycogen
    c. chlorophyll
    b. glucose
    d. fats

3. Synthesis of molecules necessary to keep cells alive is called

    a. metabolism
    c. anabolism
    b. catabolism
    d. photosynthesis

4. Which of the following is a disaccharide?

    a. sucrose
    c. fructose
    b. glucose
    d. galactose

5. Which of the following are polymers of glucose?

    a. starch
    c. glycogen
    b. cellulose
    d. all of these

6. In an alpha linkage between two glucose molecules, the oxygen atom is

    a.   pointing downward   b.   pointing upward
    c.   involved in a straight chain   d.   involved in a branched chain

7. In cellulose, glucose molecules are connected by a(n)

    a.   alpha linkage   b.   beta linkage
    c.   peptide bond   d.   amide linkage

8. In plant starch, the difference between amylose and amylopectin is

    a.   amylose is a straight chain and amylopectin is a branched chain
    b.   amylose is a branched chain and amylopectin is a straight chain
    c.   amylopectin and amylose are both straight chains with the amylose chain the longer
    d.   amylopectin and amylose are both branched chains with the amylose chain more branched

9. Which of the following are lipids?

    a.   fats and oils   b.   fatty acids
    c.   cholesterol   d.   all of these

10. Saturated fats differ from unsaturated fats in that saturated fats

    a.   have only single bonds in the fatty acid chains
    b.   have double bonds in the fatty acid chains
    c.   have triple bonds in the fatty acid chains
    d.   contain glycerol

11. Which of the following would have the highest iodine number?

    a.   monounsaturated fats   b.   diunsaturated fats
    c.   polyunsaturated fats   d.   saturated fat

12. Proteins are copolymers of

    a.   DNA   b.   RNA
    c.   amino acids   d.   enzymes

13. Amino acids form zwitterions because amino acids

    a.   have an acidic and basic end   b.   form hydrogen bonds
    c.   contain sulfur   d.   readily form covalent bonds

14. A peptide bond is a (an)

    a.   ester linkage   b.   amide linkage
    c.   alpha linkage   d.   beta linkage

15. The N-terminal of a protein is

    a. the free carboxyl end
    c. the hydroxyl end
    b. the free amino end
    d. the amide linkage

16. The primary structure of a protein is held together by

    a. ionic bonds
    c. hydrogen bonds
    b. covalent bonds
    d. disulfide bonds

17. The protein structure responsible for the folding of a protein chain is

    a. primary structure
    c. tertiary structure
    b. secondary structure
    d. quaternary structure

18. The protein structure responsible for the strength of silk and the resiliency of wool fibers is

    a. primary
    c. tertiary
    b. secondary
    d. quaternary

19. In the body, enzymes work as

    a. cofactors
    c. mRNA
    b. catalysts
    d. tRNA

20. When an enzyme reacts with a substrate

    a. the enzyme fits a section of the substrate
    b. the substrate fits the entire enzyme
    c. the substrate and enzyme both permanently change size
    d. the enzyme permanently changes shape but the substrate does not

21. Nucleic acids are composed of a chain of nucleotides containing

    a. sugars
    c. phosphate groups
    b. bases
    d. all of the above

22. The bases in nucleic acids are

    a. purines (one ring) and pyrimidines (two fused rings)
    b. purines (two fused rings) and pyrimidines (one ring)
    c. purines and pyrimidines (both one ring)
    d. purines and pyrimidines (both two fused rings)

23. Ribose is the sugar in

    a. DNA
    c. RFLP
    b. RNA
    d. cofactors

24. Double helix refers to the structure of

    a.   RNA   b.   DNA

    c.   DNA only during cell division   d.   RNA during transcription

25. Transcription is the process of transferring the genetic code from

    a.   DNA to tRNA   b.   DNA to mRNA

    c.   mRNA to tRNA   d.   mRNA to tDNA

26. A gene is a segment of a molecule of

    a.   DNA   b.   mRNA

    c.   tRNA   d.   protein

27. If a nucleic acid is completely hydrolyzed, which type of compound is not one of the products?

    a.   a base   b.   a phosphoric acid

    c.   an amino acid   d.   a sugar

28. Which set of bases does not make up a base pair usually found in nucleic acids?

    a.   adenine—thymine   b.   cytosine—guanine

    c.   uracil—thymine   d.   adenine—uracil

29. Which contains the codon?

    a.   DNA   b.   mRNA

    c.   the ribosome   d.   the protein molecule

30. Which molecule carries the anticodon?

    a.   mRNA   b.   tRNA

    c.   the ribosome   d.   the protein molecule

31. Base pairing is accomplished by

    a.   covalent bonds   b.   hydrogen bonds

    c.   ionic bonds   d.   phosphate linkages

32. When active protein synthesis is taking place in the cell, which material is not required at the ribosomes?

    a.   DNA   b.   mRNA

    c.   tRNA   d.   growing protein chain

True or False

___ 33. It is impossible for base pairing to occur in single-stranded RNA.

___ 34. DNA always incorporates equal amounts of purines and pyrimidines.

___ 35. The pairing of a purine with a pyrimidine permits the strands of a double helix to maintain a constant spacing.

___ 36. The pairing of cytosine with guanine and the pairing of adenine with thymine permit the best hydrogen bond interaction between base pairs.

___ 37. RNA contains both the anticodon triplet and the amino acid called for by the codon of mRNA.

___ 38. The codons that do not call for a specific amino acid signal the termination of protein synthesis.

___ 39. Thymine and uracil are both pyrimidines.

___ 40. Some codons call for more than one kind of amino acid.

ANSWERS

1. d	8. a	15. b	22. b	29. b	36. T
2. b	9. d	16. b	23. b	30. b	37. F
3. c	10. a	17. c	24. b	31. b	38. T
4. a	11. c	18. b	25. b	32. a	39. T
5. d	12. c	19. b	26. a	33. F	40. F
6. a	13. a	20. a	27. c	34. T	
7. b	14. b	21. d	28. c	35. T	

Food

Those Incredible Edible Chemicals

KEY TERMS

adipose tissue
aflatoxins
Agent Orange
antioxidants
arithmetic growth
arteriosclerosis
blood sugar
carbohydrates
cellulose
chain reaction
defoliation
Delaney amendment
dietary minerals
dioxins
doubling time

enrichment
essential amino acids
fat deposits
fats
food additives
free radicals
geometric growth
glycogen
GRAS list
humectant
insecticide
juvenile hormones
limiting reagent
lipoprotein
micronutrients

microprotein
organic farming
pesticides
pheromones
polyunsaturated fats
preemergent herbicides
primary plant nutrients
Rule of 72
saturated fats
secondary plant nutrients
sex attractants
starch
starvation
triglicerides
vitamins

CHAPTER SUMMARY

16.1   Carbohydrates: The Preferred Fuels
   A. Carbohydrates include sugars and starches.
   B. Sweet Chemicals: The Sugars
       1. Monosaccharides.
           a. Glucose (dextrose), called blood sugar, is used by cells to provide energy.
           b. Fructose, called fruit sugar, is found in fruit or made from glucose with the help of enzymes.
               i.   Fructose is sweeter than glucose.
               ii. Glucose in corn syrup is converted to fructose by the use of enzymes.
       2. Disaccharides.
           a. Sucrose (table sugar).
           b. Lactose (milk sugar).
       3. Americans consume 61 kg of sugar per year mostly in the form of highly processed foods and soft drinks.
   C. Digestion and metabolism of carbohydrates
       1. Glucose and fructose are absorbed directly into the blood stream from the digestive tract.
       2. Sucrose and lactose are hydrolyzed with the help of enzymes to form simpler sugars.

   a. Lactose intolerance is a condition in which a person lacks the correct enzyme to break down lactose.
  3. All monosaccharides are converted to glucose during metabolism.
   a. Galactosemia is a condition due to a deficiency of the enzyme that converts the monosaccharide galactose to glucose.
 D. Complex carbohydrates: starches
  1. Starch is hydrolyzed to glucose during digestion.
   a. More than 50 chemical reactions are needed to produce CO_2, water, and energy from starch. This is the reverse of photosynthesis.
  2. Carbohydrates supply 4 kcal of energy per gram.
   a. Excess carbohydrate is stored as glycogen and fat.
   b. Carbohydrates are the body's preferred fuel.

16.2 Fats: Energy Reserves and Cholesterol and Cardiovascular Disease
 A. Digestion and metabolism of fats
  1. Fats are digested with the help of enzymes called lipases.
  2. The products of digestion are assembled as triglycerides (esters of glycerol and fatty acid).
  3. Fats are stored in adipose tissue to serve as a protective cushion and insulation.
  4. Fats are hydrolyzed to glycerol and fatty acids. Glycerol can be "burned" to produce energy.
 B. Fats, cholesterol, and human health
  1. Saturated fats and cholesterol are associated with arteriosclerosis, which can result in heart attack and strokes.
  2. Saturated fats have a large proportion of saturated fatty acids (palmitic and stearic acid).
   a. Unsaturated fats (oils) have a large proportion of unsaturated fatty acids (linoleic and linolenic acids).
  3. Lipoprotein is a complex of cholesterol or fat with protein.
   a. Very-low-density lipoproteins (VLDL) transport triglycerides.
   b. Low-density lipoproteins (LDL) transport cholesterol to cells for use.
   c. High-density lipoproteins (HDL) transport cholesterol to the liver for eventual excretion.
   d. High levels of LDL increase risk of heart attack and stroke.
   e. Exercise increases levels of HDL.
   f. Fish oil tends to lower cholesterol and triglyceride levels.
 C. Fats are useful, but harmful in excess.
  1. Fats yield 9 kcal of energy per gram.
  2. Fats are used as fuel, to build and maintain parts of cells, and are stored as fuel reserves.
  3. The American Heart Association recommends that no more than 30% of total calories should come from fats (with no more than 10% from saturated fats).
   a. Fats and cholesterol are found in animal products (including dairy).

16.3 Proteins: Muscle and Much More
 A. Proteins are polymers of amino acids. They are needed to make muscles, hair, and enzymes.
 B. Protein metabolism: essential amino acids
  1. The human body can synthesize 12 of the 20 amino acids. The other eight (essential amino acids) must be supplied by our diet.
  2. People whose diets fail to provide all the essential amino acids in the needed quantities may suffer malnutrition even though adequate calories are provided in diet.
  3. Protein from animal sources contains all the essential amino acids. Gelatin is the only inadequate animal protein.
 C. Protein Deficiency: Kwashiorkor
  1. Our requirement for protein is 0.8 g/kg of body weight.
   a. Kwashiorkor is a protein deficiency disease common in parts of Africa.
  2. Nutrition is very important for children.

  a. The human brain reaches nearly full size by age 2. Inadequate nutrition at this time can lead to mental and physical retardation.

D Vegetarian Diets

 1. It takes less energy to produce plant protein than animal protein.

 2. Many ethnic foods provide good protein by combining a legume (deficient in methionine) with a cereal grain (deficient in tryptophan and lysine).

 3. A strict vegetarian diet is also likely to be low in vitamin B_{12}, calcium, iron, and riboflavin. Milk, eggs, cheese, fish, and supplements can correct this deficiency.

16.4 Minerals: Important Inorganic Chemicals in Our Lives

 A. A variety of inorganic compounds and dietary minerals are necessary for proper growth and repair of body tissues.

 B. Structural molecules and macromolecules make up 99% of the atoms in the human body

 C. Minerals

  Ca: bones, teeth, blood clotting, milk formation

  P: bones, teeth, nucleic acids, cell membranes

  K: intracellular cation, muscle contraction

  S: amino acids methionine and cysteine

  Na: extracellular cation, fluid pressure

  Mg: enzyme cofactor, nerve impulses

 D. Trace minerals are needed in smaller amounts but are equally important.

 1. Fe: hemoglobin for O_2 transport

 2. Iodine: proper functioning of thyroid

16.5 The Vitamins: Vital, but Not All Are Amines

 A. Vitamins are organic substances that our bodies need for good health but cannot synthesize; they must be included in the diet.

 B. Some vitamins were discovered early because of vitamin-deficiency diseases such as scurvy (vitamin C deficiency) and beriberi (thiamine or vitamin B_1 deficiency). The first such compounds characterized were amines, hence the name "vitamin."

 C. Vitamins are divided into two broad categories.

 1. Fat-soluble vitamins: A, D, E, K.

 2. Water-soluble vitamins: B complex and C.

 D. Large doses of fat-soluble vitamins can be toxic; excess water-soluble vitamins are simply excreted.

 1. Vitamins E and K are fat soluble but metabolized (not stored) and excreted.

 E. The body can store fat-soluble vitamins for future use; water-soluble vitamins are needed almost daily.

16.6 Other Essentials

 A. Dietary Fiber

 1. Insoluble fiber is a protection against colon cancer.

  a. Cellulose molecules in high-fiber foods form hydrogen bonds to water molecules.

   i. Cellulose is not digested, so it passes into the colon with its attached water. This leads to frequent and robust bowel movements.

  b. People in countries where diets are high in fiber have a lower incidence of colon cancer.

  c. Bacteria in the colon produce mutagenic substances from fecal material. Frequent bowel movements keep these mutagens flushed out.

 B. Water

 1. We need between 1.0 L and 1.5 L of water every day in addition to the water we get from food.

 2. Carbonated soft drinks lead water as the most consumed beverage in the United States.

16.7 Starvation and Fasting
  A. A body totally deprived of food soon uses up its glycogen reserves and needs to convert to fat metabolism.
  B. Increased dependence on stored fats for energy can result in ketosis.
  C. Low-carbohydrate diets induce ketosis; diets high in protein put extra stress on the liver and kidneys as the body tries to rid itself of excess nitrogen compounds.
  D. During the early stages of a total fast, the body will also break down its own proteins to try to meet its metabolic needs and to provide glucose to the brain.
     1. This results in an emaciated appearance.
     2. Ketosis is characterized by the appearance of ketones in the urine. The ketone bodies are acetoacetic acid, ß-hydroxybutyric acid, and acetone.
     3. Two of the ketone bodies are acids. Uncontrolled ketosis can lead to acidosis and death. Oxygen deficiency and dehydration are symptoms of acidosis. Acidic blood cannot transport oxygen very well, and the kidneys eliminate a lot of fluids in an attempt to remove excess acids.
  E. Fasting does not cleanse the body. It puts increased stress on the liver.
  F. Processed Food: Less Nutrition
     1. Making white flour from wheat removes protein, minerals, vitamins, and fiber (bran).
     2. Fruit peels are rich in vitamins and fiber.
     3. Some vitamins are (partially) destroyed by heat; water-soluble vitamins are leached out and discarded in the cooking water.
     4. Over half of the typical diet in the United States consists of processed food, often high in calories but low in nutrition.

16.8 Additives to Enhance our Food
  A. Food additives may be intentionally (put in product to perform a specific function) or incidentally (accidentally) added during production.
     1. Intentional additives are used to
        a. Improve nutrition.
        b. Enhance color (appearance).
        c. Enhance flavor.
        d. Retard spoilage.
        e. Provide texture.
        f. Keep foods moist.
     2. The U.S. Food and Drug Administration (FDA) regulates food additives. Since 1958, food additives must be proved safe for the intended use.
     3. Food additives (salt, spices) have been used since ancient times, but their use has increased in recent years because
        a. The population has shifted from farms to cities.
        b. Busy people want convenience foods.
        c. Competition for sales leads to the production of more attractive and tasty foods.
     4. People express concern over "chemicals" in food, but all food is chemical (and so are we).
     5. Our bodies are about 65% oxygen (by mass). Table 16.7 in the text gives the elemental composition of the human body, but few of the substances are present as free elements.
  B. Additives that Improve Nutrition
     1. Iodine is lacking in foods in land regions; small amounts of potassium iodide (KI) are added to table salt to prevent goiter (enlargement of the thyroid gland).
     2. Vitamin B$_1$ (thiamine) is added to polished rice to prevent beriberi.
     3. Iron (ferrous carbonate) is added to flour for enrichment.
     4. Vitamin C is added to fruit drinks to match that found in real fruit juices.
     5. Vitamin D is added to milk to prevent rickets.
     6. Vitamin A is added to margarine to match that found in butter.

7. Processed foods, even those with added nutrients, seldom match the nutritional value of fresh foods because only some of the nutrients lost in processing are replaced.
C. Chemicals that Taste Good
1. Some foods (spice cake, soft drinks, sausage) depend almost entirely on additives for flavor.
2. Natural flavors include spices and substances extracted from fruits and other plant materials (vanilla extract, lemon extract).
a. Chemists analyze natural flavors. Artificial flavors match the (major) components of natural ones.
b. Imitation flavors contain fewer chemicals than natural ones.
c. Flavors generally present little hazard when used in moderation.
D. Sweeteners
1. Cyclamates were banned in the United States in 1970; they are not banned in Canada.
2. Saccharin causes cancer in laboratory animals. The FDA moved to ban it, but the U.S. Congress exempted it from the Delaney amendment. May 16, 2000, saccharin was dropped from the cancer list.
3. Aspartame, a dipeptide (the methyl ester of aspartylphenylalanine) is another low-calorie sweetener.
4. Acesulfame K is a low-calorie sweetener that can survive the high temperatures of cooking.
5. Glycerol and propylene glycol are sweet; they are used as humectants (moistening agents).
6. Sorbitol and xylitol are sweet. They have the same caloric content as sugar but don't cause tooth decay.
E. Flavor Enhancers
1. Some substances, not particularly flavorful themselves, enhance other flavors.
a. Sodium chloride (table salt) is an example.
2. Monosodium glutamate (MSG) imparts a meaty flavor to foods with only a little meat.
a. Overindulgence in foods high in MSG may cause Chinese-restaurant syndrome.
3. Foods that meet FDA standards of identity must contain certain ingredients in established proportions. They may contain other ingredients, including approved additives such as MSG.
F. Spoilage Inhibitors
1. Propionic acid, benzoic acid, sorbic acid, and salts of these acids retard spoilage by inhibiting the growth of molds.
2. Nitrite salts inhibit the growth of bacteria, including those that cause botulism.
a. Only 10% of the amount used is needed to inhibit the growth of botulism.
b. Nitrites contribute to the flavor of cured meats and help keep the lean parts pink.
c. Nitrites react with HCl and amines to form carcinogenic nitrosamines, but this reaction is inhibited by ascorbic acid (vitamin C).
3. Sulfur dioxide and sulfite salts are used as disinfectants and preservatives, as bleaching agents, and to prevent foods from turning brown upon standing. These substances cause severe allergic reactions in some people.
G. Antioxidants
1. Antioxidants prevent foods containing fats and oils from becoming rancid.
2. BHA and BHT: free radical reactions.
a. Rancidity involves the reactions of fats and oils with oxygen to form free radicals.
b. These radicals react with other fat molecules to form new free radicals in a chain reaction.
c. BHT and BHA react with the radicals, halting the chain process.
3. BHT and BHA cause allergic reactions in some people and fetal abnormalities in rats. Overall, though, BHT has been shown to increase the life span of rats.
4. Vitamin E is a natural antioxidant. Lack of vitamin E causes sterility in rats, but human diets almost always have sufficient vitamin E.
a. Its action as an antioxidant is similar to BHT.
H. Food Colors
1. Some foods are naturally colored.
a. β-carotene from carrots (provitamin A)—yellow

        b.   Beet juice—red

        c.   Grape-hull extract—blue

   2. The Delaney amendment requires the FDA to ban from foods any substance shown to be carcinogenic in humans or in laboratory animals.

        a.   Several synthetic dyes have been banned to conform with the Delaney amendment.

        b.   Synthetic food dyes are low-risk, low-benefit substances.

 I.   The GRAS List

   1. Under the 1958 food additives amendment, the FDA has established a list of long-used additives "generally recognized as safe" (GRAS).

   2. Improved instruments and better experimental design have led to the banning of some GRAS substances upon reexamination.

16.9   Poisons in Your Food

  A. Many plants and animals are poisonous: some mushrooms, rhubarb leaves (oxalic acid), puffer fish.

   1. The most poisonous substance known is the botulism toxin, produced by anaerobic bacteria in improperly canned food.

  B. Carcinogens

   1. Charcoal-broiled foods contain 3, 4-benzpyrene, a carcinogen.

   2. Cinnamon and nutmeg contain safrole, a carcinogen.

   3. Aflatoxins are produced by molds that grow on peanuts and grains.

  C. Incidental Additives include pesticide residues, insect parts, and antibiotics fed to animals.

   1. Daminozide (Alar) on apples.

   2. Aminotriazole in cranberries.

   3. Polychlorinated biphenyls (PCBs) in poultry and eggs.

   4. Antibiotics added to animal feeds persist in meat and milk.

   5. DES (now banned) has been found in meat.

16.10   A World without Food Additives

  A. Without food additives, our food supply would be reduced, less safe, and less varied.

  B. The major threats to our foods (and our health) are rodent filth, insect contamination, and harmful microorganisms, not food additives.

16.11   Plants: Sun-Powered Food-Making Machines

  A. Green plants use photosynthesis to make sugars from carbon dioxide and water. The reaction also produces oxygen, replenishing the atmosphere.

  B. Structural elements of plants (carbon, hydrogen, oxygen) are derived from air and water. Other nutrients come from the soil. Energy is supplied by the sun.

  C. In a primitive society, nearly all the energy used is renewable (human work energy, sunlight), and nearly all human energy is devoted to obtaining food.

   1. One unit of human energy might yield ten units of food energy.

  D. Primitive societies usually recycle nutrients by returning wastes to the soil.

16.12   Farming with Chemicals: Fertilizers

  A. The three primary plant nutrients are nitrogen, phosphorus, and potassium.

   1. Plants cannot use N_2 directly, but some bacteria can fix nitrogen (convert it to a combined form).

   2. Plants take up nitrogen as nitrate (NO_3^-) or ammonium ions (NH_4^+).

        a.   Manure breaks down to release these inorganic ions.

   3. Population growth in the late nineteenth century led to an increased demand for fertilizers.

  B. Fritz Haber developed a process for combining nitrogen and hydrogen to make ammonia.

  C. By oxidizing part of the ammonia molecule to nitric acid and reacting the acid with ammonia, the Germans were able to make ammonium nitrate, an explosive.

   1. Ammonium nitrate is a valuable nitrogen fertilizer.

D. Ammonia can also be combined with
   1. Carbon dioxide to make urea.
   2. Sulfuric acid to make ammonium sulfate.
E. Phosphates
   1. Calcium phosphate (in bones and rocks) reacts with sulfuric acid to form superphosphate.
      a. Most modern fertilizers result from the reaction of calcium phosphate and phosphoric acid yielding water-soluble calcium dihydrogen phosphate.
   2. In plants, phosphates are incorporated into DNA and RNA.
   3. They play a vital role in photosynthesis.
   4. The availability of phosphorus is often the limiting factor in plant growth.
   5. Phosphate reserves are limited and not renewable; phosphates are scattered and lost through use.
F. Potassium
   1. Plants use potassium as simple ion K^+.
   2. The role of potassium in plants is not well understood.
   3. Uptake of potassium ions from the soil leaves the soil acidic.
      a. Each time a potassium ion enters the root tip to maintain electrical balance, a hydronium ion enters the soil.
   4. Potassium fertilizers are usually potassium chloride (KCl), which is mined in many parts of the world.
   5. Potassium salts are a nonrenewable resource; they are scattered and lost through use.
G. Other Essential Elements
   1. Three secondary nutrients (magnesium, calcium, and sulfur) are needed in moderate amounts.
   2. Seven micronutrients (boron, copper, iron, manganese, molybdenum, zinc, and chloride) are needed in small amounts.
H. Fertilizers: A Mixed Bag
   1. Complete fertilizers contain the three main nutrients.
      a. The first number on the fertilizer represents the % Nitrogen.
      b. The second is Phosphorus calculated as % P_2O_5.
      c. The third is Potassium calculated as % K_2O.
   2. To be used by plants, nutrients must be soluble in water.
      a. Nutrients are washed into streams and lakes, where they stimulate algal blooms.
      b. Nitrates also enter the groundwater.

16.13 The War against Pests
   A. Since the earliest recorded days, insect pests have destroyed crops and spread disease.
   B. Early insecticides included lead and arsenic compounds, pyrethrum and nicotine.
      1. Only a few insects are harmful; many are beneficial.
         a. Pesticides kill both harmful and beneficial insects.
         b. Because pesticides kill living things, they should be classified as biocides.
   C. DDT: The Dream Insecticide
      1. DDT, a chlorinated hydrocarbon, was found to be effective against a variety of insects and of low toxicity to humans.
      2. DDT is easily synthesized from cheap, readily available chemicals.
         a. Chlorobenzene and chloral hydrate are warmed in the presence of sulfuric acid to produce DDT.
      3. The World Health Organization estimates that chlorinated hydrocarbon pesticides such as DDT have saved 25 million lives.
         a. In India, malaria cases were reduced from 75 million to only 5 million per year.
         b. In Sri Lanka (Ceylon), malaria cases were reduced from 2.8 million per year to only 17. When spraying was terminated, the number rose again to 1 million per year.
   D. The Decline and Fall of DDT
      1. Insect resistance: Many insects have developed a resistance to DDT; they are able to detoxify it.

a. DDT has been banned in most developed countries, but it is still used in many developing countries.
2. DDT is especially toxic to fish and birds. It interferes with calcium metabolism, making eggshells weak and easily broken.
3. DDT and its metabolites break down very slowly in the environment.
4. Biological Magnification—concentration in fatty tissues.
   a. Chlorinated hydrocarbons, such as DDT and PCBs, are nerve poisons. They dissolve in the nerve sheath and interfere with the transmission of electrical impulses along these sheaths.
5. DDT is still produced for export to countries battling malaria.

E. Organic Phosphorus Compounds
1. Chlorinated hydrocarbon pesticides have been replaced in part by organic phosphorus compounds such as malathion, parathion, and diazinon.
2. These compounds generally are more toxic to mammals but less persistent in the environment and therefore residues are seldom found in food.
   a. Malathion is an exception in that it is less toxic than DDT.

F. Carbamates
1. Carbamates such as carbaryl, carbofuran, and aldicarb are narrow-spectrum insecticides, directed against only a few insect species.
   a. Chlorinated hydrocarbons and organic phosphorus compounds are broad-spectrum insecticides; they are effective against many species.
      i. Carbaryl is especially harmful to honeybees.
2. Carbamates break down rapidly in the environment, and do not accumulate in fatty tissue.

16.14 Biological Insect Controls
A. Predator insects such as praying mantises and ladybugs can be purchased and put to work.
B. Bacteria (Dipel) are used against certain destructive moth larvae.
C. Viruses have been used against cotton bollworms, grasshoppers, and other pests.
D. Some plants are bred to be resistant to insects and fungal infections.
E. Sterilization
1. If large numbers of male insects can be sterilized and released, they can overwhelm fertile males in competition for females, drastically reducing the number of offspring.
   a. Radiation sterilization has almost eliminated the screwworm fly as a pest and has effectively controlled fruitfly infestations.
      i. The 1981 Mediterranean fruitfly infestation in California was caused in part by ineffective sterilization.
F. Pheromones: The Sex Trap
1. Pheromones are substances excreted externally to mark a trail, send an alarm, or attract a mate.
2. Chemists can isolate sex attractants, determine their structure, and make synthetic versions.
3. Sex attractants have two roles in insect control.
   a. Spread widely, the attractants confuse and disorient male insects, which detect females everywhere but can't find a real one.
   b. When used as bait in traps, the attractants can lure insects to their deaths from regular pesticides (or from other means).
4. Baited traps can be used to monitor pest population to determine optimal times for spraying with traditional pesticides.
G. Juvenile Hormones
1. Juvenile hormones control the rate of development of insect young, turning off at the appropriate time so that the insect can mature to the adult stage.
2. Chemists have isolated juvenile hormones, determined their structure, and synthesized them.
3. Juvenile hormone analogs, applied as pest control agents, keep insects in a perpetual juvenile state.
   a. That's good for control of insects such as mosquitoes that are pests in the adult state.
   b. It is not good for control of insects that are pests in the juvenile (caterpillar) state.

16.15 Herbicides and Defoliants
  A. Over half of all the pesticides produced in the United States are used to kill weeds (herbicides).
  B. 2, 4-D is widely used as a growth-regulator herbicide.
  C. 2,4, 5-T, a potent defoliant, effective against woody plants, was banned by the U.S. Environmental Protection Agency in 1985.
  D. Agent Orange, a mixture of 2,4-D and 2,4,5-T, was used in Vietnam to remove enemy cover and destroy crops.
     1. Herbicide contaminants called dioxins are thought to cause birth defects.
  E. Atrazine kills plants by stopping photosynthesis.
     1. When atrazine is used in corn fields, the corn plants deactivate atrazine and are not killed. The weeds are killed.
  F. Glyphosate, a derivative of the amino acid glycine, kills all vegetation.
  G. Paraquat: A preemergent herbicide
     1. It inhibits photosynthesis in all plants but is rapidly broken down in the soil.
     2. Is used to kill weeds before plant seedlings emerge.
  H. Calcium cyanamide defoliates cotton, making mechanical harvesting possible.
  I. The benefits are increased agricultural value and control of sources of allergies and poison ivies. The risk is an uncertain increase in the level of toxin released into the environment.

16.16 Alternative Agriculture
  A. Organic farming is carried out without synthetic fertilizers or pesticides.
     1. Organic farms use less energy, but they require more human labor than conventional farms.
  B. There is no good evidence that organically grown food is more nutritious than that grown with synthetic fertilizers.
  C. Modern agriculture is energy intensive.
     1. 10% of our nation's energy is used in agriculture.
     2. Modern agriculture makes efficient use of human energy.
        a. One U.S. farmer feeds 80 people.
        b. In our society, 10% of human energy goes into food production.
     3. Ten units of petroleum energy produce one unit of food energy.

16.17 Some Malthusian Mathematics
  A. In arithmetic growth, one increment is added in each time period.
  B. In geometric growth, the sample (population) doubles in each time period.
  C. The doubling time for a population growing geometrically can be calculated from the Rule of 72.
  D. Scientific developments such as fertilizers, pesticides, and new genetic varieties of crops have helped some nations keep pace with their rapidly increasing populations.

16.18 Can We Feed a Hungry World?
  A. At the present rate of growth, the Earth's human population will increase to 10 billion by 2027.
  B. Feeding the world today depends on new plant varieties that require the use of pesticides, fertilizers, and the use of fossil fuels to power machinery.
  C. Food production may be increased substantially through genetic engineering, but present production will have to be quadrupled to meet demand by the year 2050.

CHAPTER OBJECTIVES

(You should know that...)

1.   Carbohydrates are compounds that include sugars, starches, and cellulose.

2.   Dietary fats are primarily triglycerides and are the major reason for obesity; they are high energy foods.

3.   Low-density fats are saturated and have been linked to high levels of cholesterol.

4.   Proteins are polymers of amino acids; they are specific in purpose and are used to make muscles, hair, and enzymes.

5.   Minerals are inorganic chemicals necessary for life.

6.   Vitamins are specific organic compounds required to prevent specific diseases.

7.   Fiber and water are two other essential parts of our diet.

8.   Starvation (involuntary or "crash diets") causes the body to deplete the glycogen reserves and start using fat reserves—this leads to ketosis.

9.   In general, processed foods have reduced nutritional value because of removal of skins, heating which decomposes some vitamins, and if water is used, the loss of water-soluble vitamins and minerals.

10.   Food additives are substances other than basic foodstuffs that are present in food as a result of production, processing, packaging, or storing.

11.   Food additives are regulated by the FDA and a food additive must be proven safe before introduced to a food product. There is a GRAS list of long-used food additives "Generally Regarded as Safe."

12.   Food additives include: chemicals that improve nutrition (iodine), chemicals that taste good (ginger), artificial sweeteners (aspartame), flavor enhancers (salt), spoilage inhibitors (sorbic acid), antioxidants (BHA), and food colors (beta carotene).

13.   Foodstuffs contain naturally occurring and accidentally introduced poisons whose threat needs to be assessed on an individual basis.

14.   All food ultimately comes from plants that use photosynthesis to convert carbon dioxide and water into sugars and oxygen.

15.   The three primary nutrients in fertilizers are nitrogen, phosphorous, and potassium.

16.   Nitrate fertilizers are based upon ammonium nitrate.

17.   Phosphate fertilizers are a calcium phosphate compound mixture or an ammonium phosphate compound.

18.   Potassium fertilizers usually use KCl as the source of potassium.

19. There are three secondary nutrients (magnesium, calcium, and sulfur) and 8 micronutrients needed in small amounts.

20. The problem of insects eating crops is as old as agriculture. The solution of using pesticides is a continuing story of trying to minimize their hazards with their benefits.

21. The history of DDT is useful to understand the dilemmas in using pesticides.

22. Pesticides based upon chlorinated hydrocarbons have been largely banned; organic phosphorous compounds (Malathion®) are used but are extremely toxic, and carbamates (Sevin®) are more specific.

23. Biological controls such as insect viruses and sex hormones that interrupt the reproductive cycle are increasing in use.

24. Herbicides ("weed killers") have similar concerns and a similar history to pesticides.

25. "Organic Farming" has lower costs because of lower uses of energy and no expenses for pesticides and herbicides (and less expenses for fertilizers) but the production is lower. Organic farming also has increased risks in erosion and water pollution.

26. The issue of population growth and the earth's resources available to support that population is controversial. While science and technology have increased the resources, the population has also increased but the distribution is uneven. Ultimately there is a limit to the sustainable population.

DISCUSSION

We are now ready to apply our knowledge of bonding and of molecules to the chemicals that we eat—to food. Keep in mind that a chemical compound—regardless of where it came from or how it was made—has a constant composition, structure, and properties. With these chemical principles in mind, you can go a long way toward seeing through advertising claims, fad diets, and other food-related phenomena.

Food additives serve a variety of functions. A knowledge of chemistry can help you to understand what the additives are and how they work. Chemistry alone cannot determine whether the benefit obtained is worth the risk involved. Whether or not an additive should be used involves a value judgment. A knowledge of chemistry may help you, however, to make a more rational judgment.

EXAMPLE PROBLEMS

1. The LD_{50} of the herbicide paraquat is 150 mg/kg orally in rats. If its toxicity in humans is comparable, what is the lethal dose for a 50-kg (110-lb.) person?

$$50 \text{ kg} \quad X \quad \frac{150 \text{ mg}}{1 \text{ kg}} \quad = \quad 7500 \text{ mg}$$

2. In 1986 it was estimated that world population growth had dropped to an annual rate of 1.7%. At this rate, how long will it be before the world population has doubled? Use the Rule of 72:

$$\frac{72}{1.7} \quad = \quad 42 \text{ years}$$

ADDITIONAL PROBLEMS

1. The United States population was estimated in 1986 to be increasing at a rate of 0.8% per year (excluding illegal immigration). At this rate, when will the population have doubled?

2. The LD_{50} of pyrethrins is 1.2 g/kg orally in rats. If its toxicity in humans is similar, what is the lethal dose for a 70-kg person?

SELF-TEST

Multiple Choice

1. Which of the following yields a product other than glucose upon complete digestion?

    a. amylopectin
    c. lactose
    b. cellulose
    d. starch

2. Common table sugar is more formally described as

    a. glucose
    c. maltose
    b. lactose
    d. sucrose

3. Digestion of which compound gives glucose and fructose as products?

    a. cellulose
    c. maltose
    b. lactose
    d. sucrose

4. Blood sugar is the same as

    a. fructose
    c. glycogen
    b. glucose
    d. sucrose

5. Fructose is also known as

    a. dextrose
    c. fruit sugar
    b. milk sugar
    d. table sugar

6. Which compound cannot be metabolized by humans?

    a. amylopectin
    c. cellulose
    b. amylose
    d. glycogen

7. Animal fats and vegetable oils are

    a. amides
    c. esters
    b. ethers
    d. hydrocarbons

8. The body's principal energy reserves are in the form of

    a. alcohol
    c. vitamins
    b. carbohydrates
    d. fats

9. Which class of compounds is called the basic building blocks for our bodies?

    a. carbohydrates   b. proteins
    c. fats   d. minerals

10. Which food source fails to provide adequate amounts of all the essential amino acids?

    a. corn   b. eggs
    c. fish   d. milk

11. A good source of protein, in terms of quality, is

    a. animals   b. fish
    c. eggs   d. all of these

12. Which of the following supply the most energy per gram?

    a. carbohydrates   b. proteins
    c. fats   d. vitamins

13. Which of the following does not provide adequate protein?

    a. soybean flour   b. red meat
    c. fish   d. milk

14. Which of the following is a good source of all the essential amino acids?

    a. corn   b. fresh fruit
    c. fish   d. gelatin

15. Corn grains are usually deficient in

    a. starch   b. tryptophan and lysine
    c. methionine   d. calories

16. One can get complete protein by eating a combination of whole grain bread and

    a. butter   b. jelly
    c. mayonnaise   d. beans

17. Hill's Mountaineering, Inc., offers a food for backpacking that contains twice the energy of granola per unit weight. Which of the following could it be?

    a. pancake mix   b. dried lean beef
    c. multivitamin capsules   d. butter

18. Wheat protein is deficient in

    a. phenylalanine   b. lysine
    c. methionine   d. nitrogen

19. Ions of which metals are not known to be essential to human health?

     a. calcium   b. cobalt
     c. lead   d. sodium

20. Iodide salts are important to the

     a. oxygen transport system of the body   b. function of the thyroid gland
     c. proper development of bone and teeth   d. fluid balance in cells

21. The ions of which element are incorporated in the hemoglobin molecule?

     a. iron   b. odine
     c. calcium   d. phosphorus

22. Which vitamin is water soluble?

     a. A   b. C
     c. D   d. K

23. A "high energy" food is one that is high in

     a. vitamins   b. protein
     c. calories   d. fiber

24. The major ingredient in fruit drinks is

     a. water   b. fruit juices
     c. sugar   d. vitamin C

Matching

Match each mineral with a function.

_____25. calcium   a. prevent goiter
_____26. sodium   b. fluid balance
_____27. iron   c. enzyme cofactor
_____28. iodide   d. growth of bones and teeth
_____29. cobalt   e. oxygen transport (hemoglobin)
_____30. zinc   f. part of vitamin B_{12}

Match each vitamin with the proper deficiency disease.
_____31. vitamin C   a. beriberi
_____32. vitamin B_{12} b. hemorrhage
_____33. niacin   c. pellagra
_____34. thiamine   d. pernicious anemia
_____35. vitamin D   e. rickets
_____36. vitamin E   f. scurvy
_____37. vitamin K   g. sterility

38. The most abundant element in our bodies, by mass, is

    a. oxygen
    c. carbon

    b. nitrogen
    d. sodium

39. In terms of amounts used, the three main food additives are

    a. BHT, MSG, and sodium nitrite
    b. sugar, salt, and corn syrup
    c. FD&C Yellow No. 5, saccharin, and MSG
    d. salt, pepper, and cinnamon

40. Flour that has iron and B vitamins added is said to be

    a. a perfect food
    c. contaminated

    b. enriched
    d. whole grain

41. Which substance is used to enhance flavors?

    a. BHA
    c. MSG

    b. EDB
    d. PBB

42. Natural foods are better than processed foods in that they

    a. never contain poisonous chemicals
    b. always cost less
    c. are always aesthetically more appealing
    d. may be more nutritious because vitamins, minerals, and fibers are often lost in processing

43. Sodium propionate is added in small amounts to baked goods

    a. to prevent fats or oils from turning rancid
    b. to replace minerals lost in processing
    c. to retard mold growth
    d. as an antioxidant

44. Sugar

    a. is rich in vitamins
    c. is high in protein

    b. is rich in minerals
    d. causes dental caries

45. Which is not an antioxidant?

    a. BHT
    c. MSG

    b. BHA
    d. vitamin E

46. When rats were fed the food additive BHT, they were found to have

    a. acute poisoning
    c. increased incidence of cancer

    b. chronic poisoning
    d. longer life spans

47. Natural vitamin C differs from synthetic vitamin C in

    a. chemical properties
    b. physiological properties
    c. chemical structure
    d. that it comes from a different source

48. Snapp, Krackel, and Poppy Inc. comes out with a new corn flake that promises you 100% of your recommended dietary allowance of several vitamins. You could get the same nutritive value by eating ordinary corn flakes and

    a. strawberries
    b. sugar
    c. taking a protein supplement
    d. taking a multivitamin tablet

49. Eating too much MSG may cause

    a. cancer
    b. obesity
    c. rats to live too long
    d. Chinese-restaurant syndrome

50. Vitamin E is

    a. an antioxidant
    b. frequently missing from the diet of vegetarians
    c. approved by medical authorities for the prevention of aging
    d. approved by medical authorities for treating sterility

51. Sodium nitrite reacts with HCl and amines to form

    a. saccharin
    b. BHT
    c. nitrosamines
    d. amphetamines

52. Free radicals are formed when fats react with

    a. BHT
    b. nitrites
    c. oxygen
    d. vitamin C

53. Which substance causes severe allergic reactions in some people?

    a. propionic acid
    b. sucrose
    c. sulfur dioxide
    d. FD&C Yellow No. 5

54. Which is not a natural food color?

    a. β-carotene
    b. beet juice
    c. saffron
    d. FD&C Yellow No. 5

55. Which substance has no nutritive value?

    a. FD&C Red No. 2
    b. sorbitol
    c. sucrose
    d. thiamine

56. GRAS is the FDA's

    a. slang name for marijuana
    b. good, rich artificial stuff
    c. list of food additives generally recognized as safe
    d. general registry of antioxidant substances

57. Aflatoxin B_1, found in moldy peanuts, is

    a. an antibiotic
    c. a B vitamin

    b. an anticancer drug
    d. a carcinogen, 10 million times as potent as saccharin

58. Mangled Maize, Ltd., comes out with a new banana-flavored, presweetened cereal called "Yellow Fellow." It is most likely to be rich in

    a. vitamins, minerals, and protein
    c. food additives and calories

    b. fats and proteins
    d. unsaturated fats

59. Which contains a known carcinogen (cancer-causing agent)?

    a. white bread
    c. potato chips

    b. coffee
    d. moldy peanuts

60. The most potent poison known is

    a. DDT
    c. botulin

    b. parathion
    d. cobra venom

61. Artificial colors added to foods increase

    a. vitamin content
    c. protein content

    b. nutritive value
    d. aesthetic appeal

62. The major problem with the U.S. food supply is

    a. carcinogenic additives
    c. inadequate protein

    b. toxic additives
    d. contamination by rodents, insects, and harmful microorganisms

63. The population of China in 1980 was 1 billion. It was growing at a rate of 2.0% per year. If the present rate of growth continued, the population of China would be 2 billion in

    a. 1989
    c. 2016

    b. 2000
    d. 2050

64. A population is growing geometrically. At the end of five growth periods, it has reached half the capacity of its environment. In how many more growth periods will it reach full capacity?

    a. 1
    c. 10

    b. 5
    d. 70

65. Which of the following is not a fixed form of nitrogen?

    a. N_2

    c. KNO_3

    b. NH_3

    d. H_2NCONH_2

66. As a plant grows most of its matter comes from

    a. fertilizer

    c. clay

    b. humus in the soil

    d. carbon dioxide and water

67. The elements present in complete fertilizer are

    a. C, H, O

    c. Ca, Mg, S

    b. C, H, N

    d. N, P, K

68. One should use natural fertilizers rather than chemical fertilizers because

    a. production of chemical fertilizers involves pollution and the consumption of energy
    b. use of natural fertilizers helps maintain the humus content of the soil
    c. use of natural fertilizers on fields reduces the amount of water pollution in streams
    d. all of the above

69. Which of the following is a natural pesticide?

    a. carbaryl

    c. parathion

    b. lead arsenate

    d. pyrethrum

70. DDT is

    a. cheap

    c. persistent

    b. not very toxic to humans

    d. all of these

71. DDT reaches people principally through

    a. the air they breathe

    c. the food chain

    b. the water they drink

    d. contact with contaminated animals

72. Pyrethrum is a(n)

    a. insecticide

    c. herbicide

    b. fungicide

    d. rodenticide

73. In the 1960s, thousands of robins died because they ate

    a. berries sprayed with an herbicide
    b. earthworms that had concentrated DDT from the soil
    c. sunflower seeds sprayed with parathion
    d. corn treated with a mercury-based fungicide

74. Which of the following is the most persistent pesticide?

    a. carbaryl   b. DDT
    c. parathion   d. pyrethrum

75. Compared to DDT, parathion is

    a. less toxic, more persistent   b. less toxic, less persistent
    c. more toxic, less persistent   d. more toxic, more persistent

76. Which of the following is an organic phosphorus compound?

    a. carbaryl   b. DDT
    c. parathion   d. pyrethrum

77. Chlorinated compounds such as DDT, PCB, and dioxins are concentrated in living organisms because they are

    a. fat soluble   b. foreign to nature
    c. volatile   d. water soluble

78. Insect sex attractants are not in general use because they are too

    a. toxic   b. specific
    c. expensive   d. odorous

79. An alternative to the use of "chemical" pesticides is

    a. use of predators   b. sterilization
    c. juvenile hormones   d. all of these

80. On his father's farm, Funky used a juvenile hormone to control pests. With which insect would such action be a mistake?

    a. mosquitos   b. corn earthworms
    c. grasshoppers   d. leafhoppers

81. Which form of potatoes has the highest energy input?

    a. fresh   b. canned
    c. frozen   d. dehydrated

82. The sex attractant for the common housefly is

    a. muscone   b. an alkene
    c. cow manure   d. isopentyl acetate

83. The lethal dose of parathion is about 5 mg/kg of body weight. How much would it take to kill a 50-kg farmer?

    a. 10 mg   b. 250 mg
    c. 10 kg   d. 250 kg

84. In terms of energy use in U.S. agriculture

a. energy out equals energy in
b. energy out slightly exceeds energy in
c. energy out greatly exceeds energy in
d. energy in greatly exceeds energy out

85. Compared to organic farms, how much more energy do conventional farms use?

    a. 0.1 times   b. 0.3 times
    c. 1.5 times   d. 2.3 times

ANSWERS

Additional Problems

1. 2076   2. 84g

ANSWERS

1. c	11. d	21. a	31. f	41. c	50. a	59. d	68. d	77. a
2. d	12. c	22. b	32. d	42. d	51. c	60. c	69. d	78. c
3. d	13. a	23. c	33. c	43. c	52. c	61. d	70. d	79. d
4. b	14. c	24. a	34. a	44. d	53. c	62. d	71. c	80. b
5. c	15. b	25. d	35. e	45. c	54. d	63. c	72. a	81. d
6. c	16. d	26. b	36. g	46. d	55. a	64. a	73. b	82. b
7. c	17. d	27. e	37. b	47. d	56. c	65. a	74. b	83. b
8. d	18. b	28. a	38. a	48. d	57. d	66. d	75. c	84. d
9. b	19. c	29. f	39. b	49. d	58. c	67. d	76. c	85. d
10. a	20. b	30. c	40. b					

Household Chemicals

Helps and Hazards

KEY TERMS

anionic surfactants	emollients	micelle	sebum
antiperspirants	end note	middle note	skin protection factor
astringent	eye shadow	moisturizers	soap
bleaches	hydrophilic	mousse	sunscreen lotions
builders	hydrophobic	nonionic surfactants	surfactant
cationic surfactants	hypoallergenic cosmetics	optical brighteners	surface-active agent
colognes	keratin	paint	synthetic detergents
cosmetics	lanolin	perfumes	top note
cream	lotion	resins	waxes
deodorants depilatories	mascaramelanin	saponins	

CHAPTER SUMMARY

17.1 A History of Cleaning
  A. Saponins are soapy compounds found in the leaves of certain plants that were used by some primitive peoples for washing clothes.
  B. Plant ashes contain potassium carbonate; these alkaline ashes were used by Babylonians 4,000 years ago.
  C. Most people in the Western world seldom if ever bathed using soap until the nineteenth century.

17.2 Fat + Lye → Soap
  A. Soaps are salts of long-chain carboxylic acids.
    1. Soaps are made by reacting animal fats or vegetable oils with sodium hydroxide (lye).
    2. Early soaps often contained unreacted lye.
    3. Toilet soaps contain additives, dyes, perfumes, creams, and oils.
    4. Scouring soaps contain abrasives.
    5. Few, if any, deodorant soaps contain an active deodorant.
    6. Floating soaps are formed with air to lower their density.
    7. Potassium soaps (used in shaving creams) are softer and produce a finer lather than sodium soaps.
    8. Triethanolamine soaps are used in shampoos and other cosmetics.
  B. How Soap Works
    1. Oil and greases hold dirt to skin and fabrics.
    2. Soap has an ionic (water-soluble) end and a hydrocarbon (oil-soluble) end.
      a. Soap molecules break oils into tiny globules by sticking their hydrocarbon tails into the oil, while the ionic heads remain in the aqueous phase.
      b. The oil droplets don't coalesce due to the repulsion of the charged groups.
  C. Disadvantages of Soaps

1. Under acidic conditions, natural soaps are converted to insoluble fatty acids that precipitate as greasy scum.
2. "Hard" water contains calcium, magnesium, or iron ions, which form insoluble salts with the fatty acid anions that precipitate as "bathtub ring."

D. Water Softeners
1. Sodium carbonate (washing soda) acts in two ways.
   a. It makes the water basic so that the fatty acids won't precipitate.
   b. The carbonate ions precipitate the hard water ions and keep them from forming soap scum.
2. Trisodium phosphate acts similarly.
   a. It makes the water basic.
   b. The phosphate ions precipitate calcium and magnesium ions.
3. Water softening tanks absorb the calcium, magnesium and iron ions on polymeric materials, thus softening the water.

17.3   Synthetic Detergents
A. ABS Detergents: Nondegradable
1. Alkylbenzenesulfonate (ABS) detergents were derived from propylene, benzene, and sulfuric acid, followed by neutralization.
   a. ABS detergents worked well in acidic and hard waters, but were not degraded in nature.
   b. Foaming rivers led to their ban in the 1960s.
B. LAS Detergents: Biodegradable
1. Linear alkylsulfonates (LAS) are derived from ethylene, benzene, and sulfuric acid, followed by neutralization.
   a. LAS detergents are biodegradable.

17.4   Laundry Detergent Formulations
A. Surface active agents (surfactants) are substances that stabilize the suspension of nonpolar substances in water.
B. In addition to surfactants, modern detergent formulations contain
1. Builders: substances such as phosphates and sodium carbonate that increase the detergency of phosphates.
   a. Many locales have banned the use of phosphates.
   b. Other builders include zeolites, complex aluminosilicates that tie up the hard water ions.
C. Brighteners
1. Optical brighteners (blancophors, or colorless dyes) make clothes appear brighter by absorbing ultraviolet light (invisible) and reemitting it as blue light (visible).
2. Brighteners cause mutations in microorganisms, but their only known effect on humans is skin rashes.

17.5   Liquid Laundry Detergents
A. Soap, LAS, and ABS are all anionic detergents.
1. In anionics the nonpolar part is joined to an ionic end.
B. Liquid laundry detergents make heavy use of nonionic surfactants:
1. In nonionics, the part that contains oxygen atoms is water soluble. There is no ionic charge on the molecule.
2. Nonionic detergents readily remove oily dirt, but can't keep it suspended as well as anionic surfactants.
3. Unlike most substances, nonionic detergents are more soluble in cold than hot water.
C. Dishwashing Detergents
1. Liquid dishwashing detergents for hand dishwashing generally contain anionic or nonionic surfactants as the only active ingredients.
   a. They differ mainly in the concentration of the surfactant.

b. Other ingredients—perfumes, colors, substances purporting to soften hands, are there mainly as a basis for advertising claims.

    2. Detergents for automatic dishwashers are highly alkaline and should never be used for hand dishwashing.

17.6   Quaternary Ammonium Salts: Dead Germs and Soft Fabrics

    A. The working part of a cationic surfactant is a positive ion.
        1. Most cationics are quaternary ammonium salts; they have four alkyl groups attached to a nitrogen atom that has a positive charge.
        2. Cationic detergents are not very good for cleaning.
        3. They do have germicidal action.
            a. When combined with nonionic detergents, they are used for cleaning in the food and dairy industries.
    B. Molecules with two long hydrocarbon tails (and two smaller groups on nitrogen) act as fabric softeners.
        1. They form a film on fibers, lubricating them for flexibility and softness.

17.7   Bleaches: Whiter Whites

    A. Bleaches are oxidizing agents.
        1. Liquid laundry bleaches ("chlorine bleaches") are all 5.25% solutions of sodium hypochlorite. These solutions release chlorine rapidly and can damage fabrics.
        2. Solid chlorine bleaches (e.g.. Symclosene, a cyanurate-type bleach) release chlorine slowly and are less damaging to fabrics.
        3. Oxygen-releasing bleaches contain sodium perborate. They operate well only above 65°C.
            a. They are especially effective on resin-treated polyester-cotton fabrics.
    B. Mixing household chemicals can produce toxic substances.

17.8   All-Purpose Cleaning Products

    A. All-purpose products for use in water may include
        1. Surfactants.
        2. Sodium carbonate (washing soda).
        3. Ammonia.
        4. Solvents (to dissolve grease).
        5. Disinfectants, deodorants, etc.
    B. Household ammonia has many uses.
        1. Undiluted: loosens baked-on grease, burned-on food.
        2. Diluted: cleans glass.
        3. Mixed with detergent: removes wax from linoleum.
    C. Baking soda is
        1. A mild abrasive cleaner.
        2. An absorbant of odors in the refrigerator.
D.   Vinegar is a good grease cutter.

17.9   Special Purpose Cleaners

    A. Toilet bowl cleaners
        1. Usually acidic to dissolve calcium carbonate buildup and fungal growth.
    B. Scouring powder
        1. Contains abrasives and surfactants
            a. Can scratch surfaces
    C. Glass cleaners
        1. Volatile liquids that evaporate without leaving a residue.
    D. Drain cleaners
        1. Sodium hydroxide which reacts with water to generate heat.

          a. Heat melts the grease clogging the pipes.
      E. Oven cleaners
          1. Sodium hydroxide converts greasy deposits to soap.

17.10 Organic Solvents in the Home
       A. Solvents are used to remove paint, varnish, adhesives, waxes, etc.
       B. Solvents are also used in all-purpose cleansers.
           1. Pine oil terpenes dissolve grease and have a mild disinfectant action.
           2. Petroleum distillates dissolve grease.
       C. Most of the solvents are volatile and flammable.
           1. They cause chemical pneumonia when swallowed.
           2. They cause narcosis when inhaled at high concentrations.

17.11 Paints
       A. Paints contain pigment, binder, and solvent.
           1. Titanium oxide is used as the pigment.
           2. Binders are usually tung oil or linseed oil (oil-based paints) and polymers in water-based paints.
           3. Solvents are usually alcohol, a hydrocarbon, an ester or water.

17.12 Waxes
       A. Esters of long-chain, organic (fatty) acids with long-chained alcohols.
           1. Beeswax, carnauba wax, spermaceti wax and lanolin.

17.13 Cosmetics: Personal Care Chemicals
       A. Cosmetics are defined as articles intended to be rubbed, poured, sprinkled or sprayed on, introduced into or otherwise applied to the human body or any part thereof, for cleansing, beautifying, promoting attractiveness, or altering appearance.
           1. Soap, antiperspirants, and anti-dandruff shampoos are excluded from this definition.
       B. Skin Creams and Lotions
           1. The outer layer of skin is the epidermis. The corneal layer of the epidermis is composed of dead cells.
           2. The corneal layer is mainly keratin, a tough, fibrous protein with a moisture content of about 10%.
               a. Below 10% moisture, skin is dry and flaky.
               b. Above 10% moisture, microorganisms flourish.
               c. Sebum (skin oils) protects the skin from loss of moisture.
           3. Wind and water (especially with soap) remove sebum, leaving the skin dry. Dry skin can be treated with
               a. Lotions: emulsions of oil in water (feel cool).
                   i. Typical oils include mineral oil, petroleum jelly, and natural fats and oils.
               b. Creams: emulsions of water in oil (feel greasy).
               c. Emollients provide a protective coating on the skin to prevent loss of moisture.
               d. Moisturizers that hold moisture in skin usually contain lanolin or collagen.
           4. Sunscreen lotions protect the skin from harmful ultraviolet radiation. The most common ingredient was para-aminobenzoic acid.
               a. Skin protection factors (SPF) indicate how long a person can remain in the sun without burning.
               b. Melanin is a dark skin pigment; its formation during tanning is stimulated by long-wave ultraviolet radiation.
                   i. Short-wave ultraviolet radiation can cause skin cancer.
               c. Excessive sunbathing causes premature aging of the skin and can lead to skin cancer.
                   i. Cigarette smoking causes premature wrinkling of the skin.
       C. Lipsticks

      1. Lipsticks are composed of an oil (often castor oil) and a wax: the wax makes lipsticks firmer than creams.
  D. Eye make-up
      1. Mascara, composed of a base of fats and waxes, is used to darken eyelashes.
          a. It is colored by mineral pigments.
      2. Eye shadow is composed of a base of petroleum jelly and fats, oils, and waxes.
          a. It is colored by dyes and pigments.
  E. Deodorants and Antiperspirants
      1. Deodorants are perfumed products designed to mask body odor.
      2. Antiperspirants retard perspiration and therefore are classified as a drug.
          a. Nearly all antiperspirants have aluminum chlorohydrate as the only active ingredient.
          b. Aluminum chlorohydrate is an astringent; it constricts the openings of sweat glands.

17.14   Toothpaste: Soap with Grit and Flavor
  A. Toothpastes have two essential ingredients, a detergent and an abrasive.
  B. Sodium lauryl sulfate is a typical detergent used in toothpastes, but any pharmaceutical-grade soap or detergent would do.
  C. Other ingredients include flavors, colors, and sweeteners.
  D. Tooth decay is caused by bacteria that convert sugars to
      1. Dextrans (plaque).
      2. Acids (such as lactic acid) that dissolve tooth enamel.
  E. Fluorides harden tooth enamel, reducing the incidence of decay.
      1. Fluorides convert hydroxyapatite to fluorapatite, a harder material.
      2. Legally, fluoride toothpastes are drugs, not cosmetics.
  F. Hydrogen peroxide and baking soda are used to prevent gum disease, the major cause of adult tooth loss.

17.15   Perfumes, Colognes, and Aftershaves
  A. Perfumes are extracts from natural materials or similar materials put together by chemists.
  B. Perfumes are characterized by notes (components with differing volatility).
      1. <u>Top</u> <u>note</u>: the most volatile fraction; the first aroma detected when applied.
      2. <u>Middle</u> <u>note</u>: intermediate in volatility; responsible for the lingering aroma after the top note is gone.
      3. <u>End</u> <u>note</u>: low in volatility.
  C. Several fruity or flowery compounds are synthesized in large quantities for use in perfumes and as fragrances for commercial products.
  D. Musks are added to counteract sweet, flowery odors.
  E. Andron by Jovan claims to contain a human sex attractant.
  F. Colognes are diluted perfumes.
  G. Aftershave lotions are colognes, sometimes with menthol for a cooling effect or an emollient for soothing the scraped skin.
  H. Many people have allergic reactions to perfumes.
      1. Hypoallergenic cosmetics often omit perfumes.

17.16   Some Hairy Chemistry
  A. Hair is composed of the fibrous protein keratin.
  B. Protein molecules in hair are strongly held together by four types of forces.
      1. <u>Hydrogen</u> <u>bonds</u>—disrupted by water.
      2. <u>Salt</u> <u>bridges</u>—destroyed by changes in pH.
      3. <u>Disulfide</u> <u>linkages</u>—broken and destroyed by permanent wave and hair straightening treatments.
      4. <u>Hydrophobic</u> <u>interactions</u>.
  C. Shampoo
      1. When hair is washed, the keratin absorbs water. The water disrupts hydrogen bonds and some salt bridges.

   a. The hair is softened and made more stretchable.
  2. Hair shafts are dead; only the root is alive. The hair is lubricated by sebum.
   a. Sebum adheres dirt to hair.
   b. Washing hair removes the oil and dirt.
  3. Shampoo for adults usually has an anionic surfactant, such as sodium dodecyl sulfate, as the principal active ingredient.
  4. Baby shampoos have an amphoteric surfactant with both a negatively charged oxygen and a positively charged nitrogen.
  5. Most components other than a detergent are in a shampoo only as the basis for advertising claims.
  6. Hair is protein with both acidic and basic groups.
   a. Most shampoos have a pH between 5 and 8 which does not damage hair or scalp.
   b. Protein shampoos condition hair by coating the hair shaft with protein (glue).
   c. Shampoos for dry or oily hair differ only in the relative amounts of detergent.
 D. Hair Coloring
  1. Hair (and skin) are colored by pigments.
   a. Melanin is responsible for brown and black colors.
   b. Phaeomelanin is the pigment in red hair.
   c. Blondes have little of either pigment; brunettes can become blondes by oxidizing the pigments with hydrogen peroxide.
  2. Permanent hair dyes often are derivatives of para-phenylenediamine. These compounds penetrate the hair shaft and are oxidized to colored products (presumably quinones).
   a. *para*-phenylenediamine produces a black color.
   b. *para*-aminodiphenylamine-sulfonic acid is used for blondes.
   c. Intermediate colors use other derivatives.
   d. Several of the hair-coloring diamines have been shown to be carcinogenic or mutagenic.
  3. Hair treatments that restore color gradually use lead acetate solutions.
   a. The lead ions penetrate the hair and react with sulfur to form black, insoluble lead sulfide.
 E. Permanent Waving: Chemistry to Curl Your Hair
  1. Adjacent protein molecules in hair are cross-linked by disulfide groups. To put curl in hair
   a. A reducing agent is used to rupture the disulfide linkages.
   b. The hair is set on curlers; the protein chains slide in relation to one another.
   c. Disulfide linkages are formed in new positions.
   d. The same chemical process can be used to straighten hair.
 F. Hair Sprays
  1. Hair can be held in place by resins (often polyvinylpyrrolidone or its copolymers).
   a. The resin can be dissolved in a solvent and applied as a spray.
  2. Holding resins can also be formulated as mousses (foams or froths).
 G. Hair Removers (depilatories)
  1. Strongly basic sulfur compounds that destroy peptide bonds.
  2. They can damage skin.
 H. Hair Restorers
  1. Minoxidil dilates blood vessels and produces growth of fine hair to skin containing hair follicles.

17.7   Well-informed Consumer
    A. Most cosmetics are formulated from inexpensive ingredients.
    B. You don't have to pay a lot for extra ingredients that contribute little to the performance of the product.

CHAPTER OBJECTIVES

(You should know that...)

1. We use many chemicals in the home and some of them are hazardous.

2. Cleaning with soap is a modern development.

3. Animal fat treated with NaOH forms natural soap. This is a long hydrocarbon chain that dissolves oils and a water soluble salt of a carboxylic acid.

4. Na salt soaps are precipitated by "hard water,",K salt soaps are softer and used as liquid soaps; Ammonium salt soaps are used in shampoos.

5. Water softeners remove Ca, Mg, and Fe ions.

6. Synthetic detergents were developed after World War II to avoid the problems of hard water. The ABS detergents were non-biodegradeable and discontinued because of ecological problems and replaced with LAS detergents.

7. Laundry detergents have several components: builders that improve their cleaning ability and brighteners that absorb UV light and re-emit in the visible range making them look cleaner.

8. Liquid detergents are not as effective cleaners but more convenient than solid detergents. These are mostly anionic surfactants: the polar end is negative.

9. Quaternary ammonium salts are cationic surfactants (the polar end is positive) and are useful in killing germs.

10. Bleaches are oxidizing agents and include NaOCl (a chlorine bleach) and borax.

11. Some general purpose cleaners are ammonia (good for baked on food and grease but hard on aluminum materials) and baking soda (a good mild abrasive and odor absorber) and vinegar (good for grease but hard on marble).

12. Special purpose cleaners include toilet bowl cleaners (dissolvers of limestone or $CaCO_3$ deposits such as HCl or sodium bisulfate); scouring powder (silica) glass cleaners (isopropyl alcohol or ammonia); drain cleaners (grease active) chemicals such as NaOH, and oven cleaners (NaOH to dissolve the grease).

13. There are a number of organic solvents (gasoline) that are highly flammable, toxic when swallowed, and narcotic at high concentrations–a problem with people "sniffing glue" and other household solvents and dying.

14. Paints are composed of a pigment, a binder, and a solvent.

15. Waxes are esters of long-chain organic acids and long chain alcohols. They are used to form protective coatings.

16. There are a variety of cosmetic products defined legally as articles "for cleansing, beautifying, promoting attractiveness or altering the appearance..." but soaps are specifically excluded.

17. Cosmetic products include: skin creams and lotions (can form a protective coating, soften skin, hold in moisture or block UV radiation); lipstick (a dye with some protective properties); eye make-up (pigments with a wax base); and deodorants and antiperspirants (perfume to mask body odor and germicide to kill bacteria).

18. Toothpastes contain soaps and abrasives. Stannous fluoride is sometimes added to make the enamel stronger and more resistant to decay.

19. Perfumes, colognes, and aftershaves are complex compounds with attractive smells dissolved in alcohol and water.

20. Hair is primarily fibrous keratin, a protein but with many more disulfide linkages than the keratin of skin.

21. Shampoos use a synthetic detergent as a cleansing agent with many shampoos having additives such as proteins (gives more body); conditioners that soften the hair like fabric softeners; different amount of cleansers for oily, normal and dry hair; and "natural" additives such as honey, herbs, etc., which have no evidence of effectiveness.

22. The amount of melanin (brown-black pigment) and phaemelanin determines the color of hair. Bleaches oxidize these pigments and hair colorants can be added. Hair treatments to remove "grayness" use lead compounds to form lead sulfide with the hair.

23. Permanents break and reform the hydrogen bonds in the proteins of hair.

24. Hair sprays are solid or semisolid organic compounds that form a sticky layer on the hair.

25. Hair removers (depilatories) are strong bases that destroy the peptides in hair so that it can be washed off.

26. Hair restorers (Rogaine®) dilates blood vessels in hair.

DISCUSSION

In this chapter we apply some of the principles learned earlier to a study of some of the chemicals used in and around the home. Our main focus is on chemicals used in cleaning, for these are among the most common—and often the most dangerous—of the household chemicals. Others are discussed elsewhere. Pesticides, fertilizers, and other "farm" chemicals (discussed in Chapter 16) are also used on lawns, gardens, and on household plants.

In this chapter, we also apply our knowledge of chemistry to the substances we put on our skin and hair to make us look or smell better. Keep in mind that a particular chemical compound has its own characteristic set of properties. These are exhibited at all times. Some of the properties are desirable; often some are not. The properties are independent of our wishes. Often they fall far short of extravagant advertising claims. With the introduction to cosmetics provided in Chapter 17 in the text, you will be better equipped to judge for yourself the validity of some of the assertions of advertisers.

SELF-TEST

Multiple Choice

1. The salts of long-chain carboxylic acids are called

   a. soaps
   c. fabric softeners
   b. synthetic detergents
   d. bleaches

2. Which of these is true of soap?

   a. soaps that are biodegradable increase the BOD of water
   b. soap is an excellent cleanser in soft water
   c. soap works poorly in hard water
   d. all of the above are true

3. Soap was made in the second century by which of these processes?

   a. boiling saponins with potassium carbonate
   b. boiling animal fats with lye
   c. boiling potassium carbonate and oils
   d. mixing oils and acids

4. Which of these is true of the action of soap in removing dirt?

   a. the hydrocarbon end of soap is soluble in water
   b. the ionic end is soluble in oils
   c. one end of soap is soluble in water and the other in oil
   d. all of the above

5. Water alone is not as effective as it is with soap because

   a. oil and dirt are not soluble in water
   b. water and oil are both quite polar
   c. soap is soluble in water and oil at the same time
   d. all of the above

6. Which compound would not be expected to exhibit detergent action?

   a. $CH_3CH_2CH_2CH_2CH_2CH_2CH_2CH_2CH_2CH_2CH_2CH_2CH_2CH_2CH_2COO^-Na^+$
   b. $CH_3CH_2CH_2CH_2CH_2CH_2CH_2CH_2CH_2CH_2CH_2CH_2CH_2CH_2CH_2COOH$
   c. $CH_3CH_2CH_2CH_2CH_2CH_2CH_2CH_2CH_2CH_2CH_2CH_2CH_2CH_2CH_2OSO_3^-NA^+$
   d. $CH_3CH_2CH_2CH_2CH_2CH_2CH_2CH_2CH_2CH_2CH_2CH_2N^+Me_3Cl^-$

7. Bathtub ring results from the presence in water of

   a. Na^+
   c. grease
   b. Ca^{2+}
   d. phosphates

8. Increased interest in cleanliness resulted from the discovery of

    a. soap   b. phosphates
    c. disease-causing microorganisms   d. America

9. Floating soaps differ from other soaps in that they contain

    a. air   b. oil
    c. potassium   d. helium

10. In acidic solutions, soaps are converted into

    a. bases   b. phenols
    c. esters   d. carboxylic acids

11. Synthetic detergents, called ABS, replaced soaps in the early 1960s. The biggest disadvantage of ABS was that they

    a. failed to work in hard water
    b. precipitated under acidic conditions
    c. were not broken down by microorganisms
    d. did not work in alkaline water

12. The foaming rivers of the early 1960s resulted from

    a. synthetic detergents
    b. an increased population of soap users
    c. an increased BOD from biodegradable soaps
    d. the use of NTA

13. Most bleaches contain the element

    a. phosphorus   b. nitrogen
    c. potassium   d. chlorine

14. Oxygen-releasing bleaches usually contain

    a. NaOCl   b. cyanurates
    c. phosphates   d. sodium perborate

15. The only active ingredient in most household liquid bleaches is

    a. sodium hydroxide   b. sodium bicarbonate
    c. sodium hypochlorite   d. chlorine

16. Of the following, which substance is not used as a water-softening agent?

    a. sodium carbonate   b. sodium hydroxide
    c. sodium zeolites   d. trisodium phosphate

17. Ions of which element do not cause hard water?

   a. calcium
   b. iron
   c. magnesium
   d. potassium

18. Which carboxylic acid salt is insoluble in water? (R stands for a long-chain alkyl group.)

   a. $(RCOO)_2Ca$
   b. RCOOK
   c. RCOONa
   d. all are soluble

19. Substances that absorb ultraviolet light and re-emit it as visible light are called

   a. bleaches
   b. fabric softeners
   c. optical brighteners
   d. water softeners

20. Petroleum distillates are

   a. flammable
   b. grease cutters
   c. narcotic at high concentrations
   d. all of these

Matching

Caution: More than one response may be needed for some of the items.

___21. NaOCl

___22. 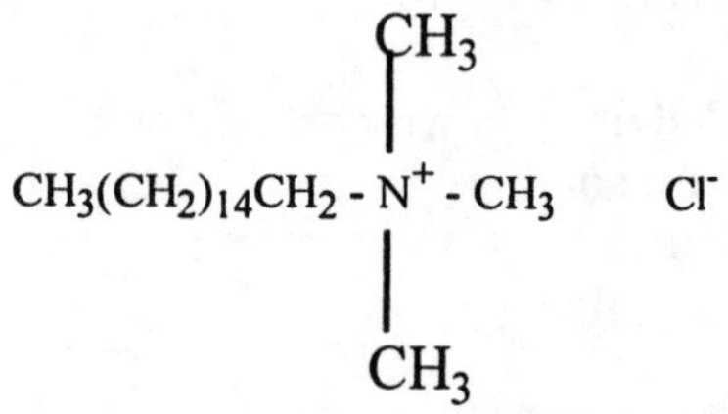 $CH_3(CH_2)_{14}CH_2 - \overset{\overset{\displaystyle CH_3}{|}}{\underset{\underset{\displaystyle CH_3}{|}}{N^+}} - CH_3$ Cl^-

___23. 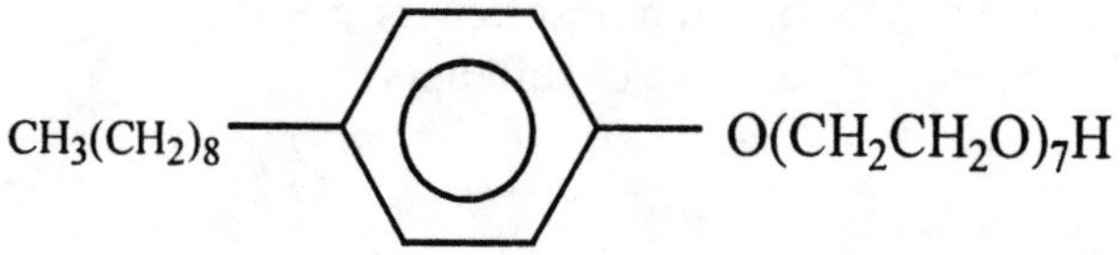 $CH_3(CH_2)_8 -$⟨◯⟩$- O(CH_2CH_2O)_7H$

___24. $CH_3(CH_2)_{16}COO^-Na^+$

___25. 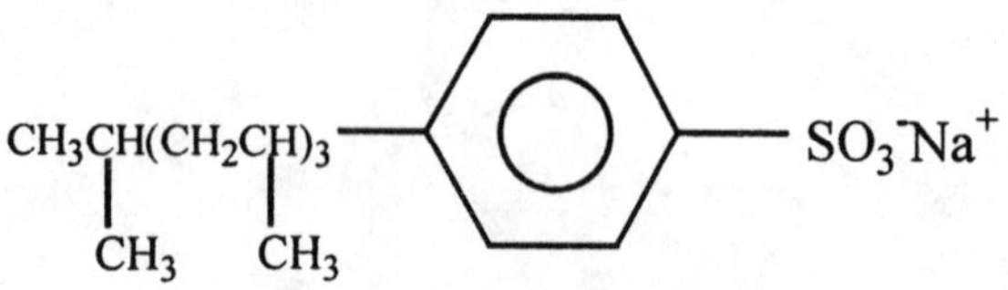 $CH_3\underset{\underset{\displaystyle CH_3}{|}}{CH}(CH_2\underset{\underset{\displaystyle CH_3}{|}}{CH})_3 -$⟨◯⟩$- SO_3^-Na^+$

___26. 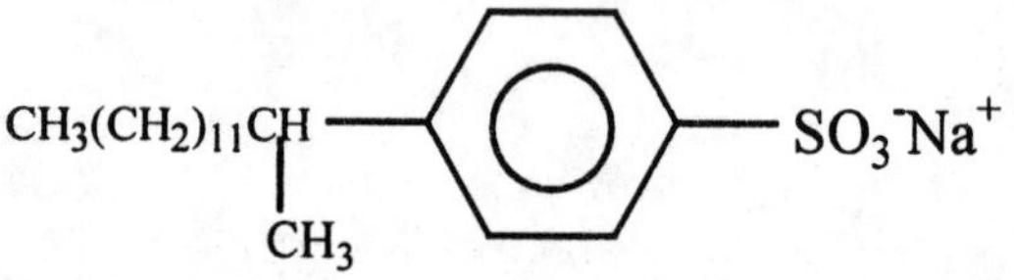 $CH_3(CH_2)_{11}\underset{\underset{\displaystyle CH_3}{|}}{CH} -$⟨◯⟩$- SO_3^-Na^+$

___27. $NaBO_2 \cdot H_2O_2$

___28.

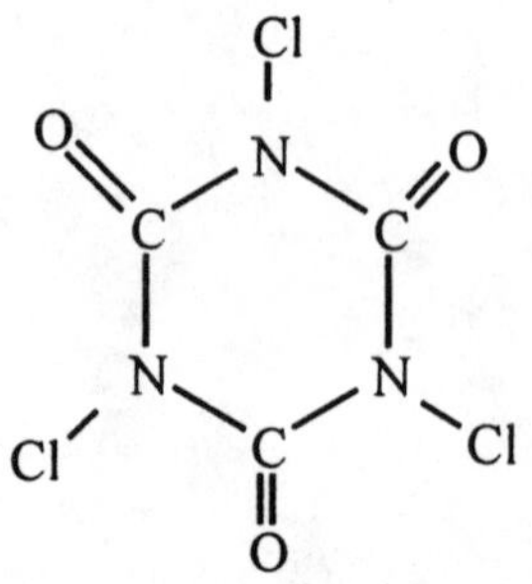

___29. 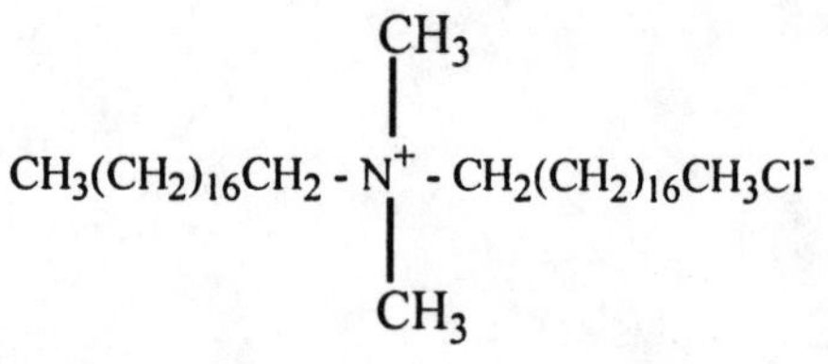 $CH_3(CH_2)_{16}CH_2 - \overset{\overset{\displaystyle CH_3}{|}}{\underset{\underset{\displaystyle CH_3}{|}}{N^+}} - CH_2(CH_2)_{16}CH_3Cl^-$

a. cyanurate-type bleach
b. germicide
c. soap
d. fabric softner
e. oxygen-releasing bleach
f. nonionic detergent
g. ABS
h. active ingredient in liquid bleaches
i. cationic detergent
k. LAS
l. nonbiodegradable detergent

30. Under U.S. law, which is not a cosmetic?

    a. cologne
    c. shampoo

    b. lipstick
    d. soap

31. The two main ingredients in most brands of toothpaste are detergent and a(n)

    a. buffer for pH balance
    c. fruit flavor

    b. sweetener
    d. abrasive

32. Fluoride in drinking water or toothpaste retards tooth decay by

    a. killing decay-causing bacteria
    b. strengthening tooth enamel
    c. deactivating salivary enzymes
    d. stopping carbohydrate metabolism

33. The main cause of tooth loss in adults in the United States is

    a. accidents
    c. gum disease

    b. fluorides
    d. sugars

34. Which is a likely major ingredient of lotions and creams?

    a. an abrasive
    c. petroleum jelly or mineral oil

    b. a detergent
    d. aluminum chlorohydrate

35. Emollients act

    a. by releasing water to keep the skin wet
    b. as detergents
    c. by forming a protective layer on the skin
    d. by retarding perspiration

36. An emulsion of water in oil is called a

    a. cream
    c. lotion

    b. lipstick
    d. moisturizer

37. The two main ingredients in most lipsticks are a wax and

    a. aluminum chlorohydrate
    c. lanolin

    b. castor oil
    d. a detergent

38. For a long time para-aminobenzoic acid (PABA) was the main ingredient in

    a. face creams
    c. sunscreen lotions

    b. mascara
    d. toothpastes

39. In suntanning, long-wave ultraviolet rays promote the formation of

    a.   collagen   b.   keratin
    c.   melanin   d.   sebum

40. Lanolin is

    a.   a hydrocarbon oil   b.   a hydrocarbon wax
    c.   a plant wax   d.   wax from sheep's wool

41. Aftershave lotions are 60 to 75% aqueous

    a.   acetic acid   b.   alcohol
    c.   mineral oil   d.   castor oil

42. The cooling effect of some aftershave lotions is provided by

    a.   ice crystals   b.   ether
    c.   menthol   d.   cologne

43. The most volatile components of a perfume are called

    a.   the end note   b.   colognes
    c.   musks   d.   the top note

44. Antiperspirants are drugs because they

    a.   change a normal body function   b.   contain narcotic propellants
    c.   contain perfume   d.   contain oxidizing agents

45. Aluminum chlorohydrate is the active ingredient in nearly all

    a.   antiperspirants   b.   lipsticks
    c.   skin lotions   d.   deodorants

46. Herbal essence, added to shampoo,

    a.   gives it pH balance   b.   feeds hair roots
    c.   gives hair body   d.   attracts insects

47. Which of the following is a drug?

    a.   deodorant   b.   antidandruff shampoo
    c.   baby shampoo   d.   lanolin

48. Strawberry essence, added to shampoo,

    a.   gives it pH balance   b.   feeds hair roots
    c.   gives hair body   d.   attracts insects

49. Hair is

    a. protein   b. carbohydrate
    c. fat   d. cellulose

50. The principal ingredient of any shampoo is

    a. a detergent   b. protein
    c. an oil   d. a buffer to balance pH

51. Which of the following can nourish hair?

    a. beer   b. wheat germ
    c. vitamins   d. none of these

52. What type of detergent is found in nearly all baby shampoos?

    a. cationic   b. nonionic
    c. amphoteric   d. none of these

53. Shampoos for adults usually employ

    a. soap   b. anionic detergents
    c. cationic detergents   d. nonionic detergents

54. In shampoos, pH balance means

    a. equal amounts of phosphorus and hydrogen
    b. equal numbers of protons and hydrogen ions
    c. a pH of exactly 7
    d. a pH range that will not harm eyes or skin

55. Most conditioners for hair are

    a. detergents   b. resins
    c. proteins   d. herbal essences

56. Bleaches for hair oxidize colored pigments to

    a. carbon dioxide and water   b. hydrogen peroxide
    c. disulfide linkages   d. colorless compounds

57. Most hair dyes are colorless compounds that are absorbed into hair and then oxidized to

    a. carbon dioxide and water   b. hydrogen peroxide
    c. para-phenylenediamine   d. colored compounds

58. Curling or straightening of hair involves breaking and reforming of

    a. protein chains   b. hair fibers
    c. disulfide linkages   d. diamine bonds

59. Hair sprays are principally

    a. oils   b. waxes
    c. synthetic resins   d. proteins

60. The color of hair that is treated to establish color gradually is due to

    a. diamines   b. lead sulfide
    c. peroxides   d. quinones

ANSWERS

1. a	11. c	21. h	31. d	41. b	51. d
2. d	12. a	22. b, i	32. b	42. c	52. c
3. b	13. d	23. f	33. c	43. d	53. b
4. c	14. d	24. c	34. c	44. a	54. d
5. c	15. c	25. i, g	35. c	45. a	55. c
6. b	16. b	26. i, k	36. a	46. d	56. d
7. b	17. d	27. e	37. b	47. b	57. d
8. c	18. a	28. a	38. c	48. d	58. c
9. a	19. c	29. d	39. c	49. a	59. c
10. d	20. d	30. d	40. d	50. a	60. b

Fitness and Health
Some Chemical Connections

KEY TERMS

aerobic exercise
anabolic steroids
anaerobic exercise
diuretic
electrolytes
endorphins

fast-twitch fibers
glycogen
heat stroke
lactate threshold
neurotrophies
oxygen debt

Recommended Dietary Allowance (RDA)
restorative drugs
set-point theory
slow-twitch fibers
training effect

CHAPTER SUMMARY

18.1 Nutrition
   A. Good health requires a nutritious diet.
   B. Fewer calories
      1. Overeating is a serious problem in the United States.
   C. Less fat but more starch
      1. Americans consume too much fat in their diet.
   D. Less protein, especially red meat
      1. With high-protein diets, weight gain is faster and weight loss is more difficult.
   E. Nutrition and the Athlete
      1. The only thing athletes need beyond what is needed by sedentary individuals is calories to fuel their activities. These are best supplied by an increase in carbohydrate intake.
      2. Athletes do not need extra protein in their diets.
         a. Muscles are built through exercise, not from eating protein.
         b. Protein metabolism produces toxic wastes that tax the liver and kidneys.
      3. If exercise is ended, the muscles shrink in about 2 days and are completely gone in about 2 months.
         a. Muscle is protein: it does not "turn into fat."
      4. When muscle contracts against resistance, creatine (an amino acid) is released.
         a. Creatine stimulates production of protein (myosin) thus building muscle.

18.2 Vitamins and Minerals
   A. There are times when the body has an increased demand for vitamins and minerals.
      1. These include pregnancy, rapid growth, recovery from disease, and trauma.
   B. Recommended Daily Allowance (RDA) are values set by the FDA for the amounts of vitamins and minerals needed to prevent deficiency diseases.
   C. Vitamin A
      1. Essential for good vision, bone development, and skin maintenance.

2. May confer resistance to some cancers.
3. Beta carotene is a precursor to Vitamin A.
4. Large doses can be toxic.
   D. Vitamin B complex
1. Eight members of the Vitamin B family
   a. Important in maintaining skin and nervous system.
2. B_6 (pyridoxine) reduces swelling and pain in carpal tunnel syndrome.
   a. Large doses of B_6 can be toxic.
3. B_3 (Niacin) offers relief from arthritis and rheumatism.
4. B_{12} deficiency can lead to pernicious anemia.
   E. Linus Pauling, Vitamin C, and the Common Cold
1. Linus Pauling recommends daily doses of Vitamin C much higher than the RDA, specifically 250 to 15,000 mg to prevent the common cold.
   a. The RDA for Vitamin C is 60 mg/day to prevent scurvy.
2. This amount of Vitamin C may also be used to prevent and/or treat other viruses.
3. Vitamin C has already been shown to help
   a. Heal wounds and gastric ulcers.
   b. Increase the body's production of interferons (part of the immune system).
   F. Vitamin D
1. Steroid-type vitamin promotes absorption of calcium and phosphorus to maintain bones.
   G. Vitamin E
1. Useful in maintaining cardiovascular system.
2. Low levels associated with sterility and muscular dystrophy.
3. Potent antioxidant that can aid in preventing some physiological damage from aging.
4. Vitamin A is oxidized to inactive form when Vitamin E is not present.

18.3 Body Fluids: Electrolytes
   A. We need a balance of fluids and electrolytes. The principal electrolytes needed for proper cellular function are sodium ions, potassium ions, and chloride ions.
   B. Our need for fluids is best met by drinking plain water.
1. Caffeine and alcohol are diuretics; beverages containing these drugs actually deplete the body of water.
2. Commercial thirst-quenching drinks, designed to replace electrolytes lost through sweating, are marginal at best; they can actually draw water from the tissues into the digestive tract resulting in diarrhea.
   C. Thirst is a delayed reaction to water loss.
1. Cloudy, yellow urine indicates dehydration.
   D. Severe dehydration can lead to heat stroke and (unless treated promptly) death.
   E. Thirst-quencher drinks are designed to replace salts lost through sweating. They are of marginal value.

18.4 Weight-Loss Diets
   A. Most quick weight-loss diets rely on a gimmick.
1. Many use a diuretic to increase urine output. Weight loss is water loss, quickly regained when the body is rehydrated.
2. Other diets, low in carbohydrates, depend on glycogen depletion.
   a. Glycogen molecules have many OH groups that hang onto water molecules through hydrogen bonding. Each pound of glycogen carries about 3 lb of water. Depleting the pound of glycogen causes a weight loss of 4 lb.
   b. The weight is quickly regained when carbohydrates are returned to the diet.
   B. The most fat you can lose in a day, even with a total fast, is about 2/3 lb.
1. The body won't burn just fat; if carbohydrates are not supplied in the diet, the body will break down muscle tissue to make glucose.

2. Weight loss through dieting includes loss of muscle mass as well as fat; weight regained (without exercise) is pure fat.
C. The most sensible weight-loss program is to follow a balanced diet (reduced somewhat in calories) and engage in a reasonable exercise program.
D. Weight lost through dieting is quickly regained when the dieter resumes old eating habits.
1. According to the set-point theory, each of us has a unique level of circulating fatty acids below which the hypothalamus triggers the hunger mechanism.
2. The set-point can be lowered by exercise.

18.5   Diet and Exercise
A. Millions of Americans diet for weight loss.
1. Many popular diet plans are grossly unbalanced and tend to be deficient in iron, calcium, and potassium.
2. Nutritional deficiencies, decreased resistance to disease, a decline in general health, and even death have resulted from some of the extreme diets.
B. Weight loss or gain is based on the law of conservation of energy.
1. If we take in more calories than we use up, the excess is stored as fat.
2. If we take in fewer calories than we need for our activities, our bodies burn some of the stored fat.
3. One pound of adipose (fatty) tissue contains 3500 calories and requires 200 miles of blood capillaries to serve its cells.

18.6   Measuring Fitness
A. Defined as 700 x body wt (lb)/height (in).
B. Values between 20 and 25 are average.
C. Values above 27 are unhealthy.

18.7   Some Chemistry of Muscles
A. Some of the energy from glucose or fatty acid metabolism is stored as adenosine triphosphate (ATP).
1. Muscle contains the proteins actin and myosin in a loose complex called actomyosin.
   a. When ATP is added to isolated actomyosin, the muscle fibers contract, which implies that ATP is the energy source for muscle contraction.
B. Aerobic exercise: plenty of oxygen.
1. When muscle contraction begins, glycogen is converted to pyruvic acid.
2. If sufficient oxygen is present (as in aerobic exercise), the pyruvic acid is oxidized to carbon dioxide and water.
C. Anaerobic exercise and oxygen debt.
1. If sufficient oxygen is not available (as in anaerobic exercise), pyruvic acid is reduced to lactic acid.
2. This lactic acid buildup leads to a pH drop and deactivation of muscle enzymes, described as "muscle fatigue."
3. The overworked muscles are incurring an oxygen debt that needs to be repaid after the strenuous exercise is over.
4. When glycogen stores are depleted, muscle cells can switch to fat metabolism.
   a. Fats are the main source of energy for sustained activity of low to moderate intensity.
D. Muscle Fibers: Twitch Kind Do You Have?
1. There are two classes of muscle fibers
   a. Fast-twitch (Type IIB) for anaerobic activity.
   b. Slow-twitch (Type I) for aerobic activity.
2. Slow-twitch fibers (Type I): endurance activities.
   a. Best suited for aerobic work of light or moderate intensity for sustained periods of time.
   b. High respiratory capacity and myoglobin levels of slow-twitch fibers help supply oxygen for sustained exercise, like long-distance running.

    c. Slow-twitch fibers have a low ability to hydrolyze glycogen and low actomyosin catalytic activity.
3. Fast-twitch fibers (type IIB) burst of power.
    a. Best suited for anaerobic short bursts.
    b. Low respiratory capacity and myoglobin levels of fast-twitch fibers are designed for quick bursts of energy like sprints.
    c. Fast-twitch fibers have high capacity for glycogen hydrolysis and high catalytic activity of actomyosin, which facilitate rapid ATP production and ability to hydrolyze ATP quickly.
4. Building muscles.
    a. Endurance training increases myoglobin levels in muscles (slow-twitch).
    b. Weight training develops fast-twitch muscles. Muscles increase in size.

18.8 Drugs and the Athlete
  A. Restorative drugs are used to remedy the effects of performance—pain, soreness, and injury.
    1. Painkillers include aspirin, acetaminophen, ibuprofen, and methyl salicylate.
    2. Anti-inflammatory drugs include aspirin and ibuprofen and cortisone derivatives.
  B. Stimulant drugs
    1. Caffeine could conserve glycogen, speed heart rate, and increase metabolism, but the effect is small.
    2. Cocaine, like amphetamines, stimulates the central nervous system and increases alertness and muscle tension. It can also mask fatigue and give a sense of increased stamina.
  C. Anabolic steroids
    1. Anabolic steroid hormones seem to increase muscle mass.
    2. Side effects in males include testicular atrophy and loss of function, impotence, acne, liver damage (including liver cancer), edema, elevated cholesterol levels, and growth of breasts.
    3. Anabolic steroids (derived from male sex hormones) make women more masculine. In addition to muscles, they develop baldness, extra body and facial hair, a deepened voice, and menstrual irregularities.
  D. Drugs, athletic performance, and drug screening
    1. Use of drugs to increase athletic performance is illegal and physically dangerous.
      a. Stimulants give a short lived and false sense of confidence.
      b. Following the high of stimulants is extreme depression.
    2. Blood and urine screening for illegal drugs is becoming standard practice.

18.9 Exercise and the Brain
  A. Vigorous exercise causes the body to produce its own painkillers, endorphins.
  B. These substances can produce euphoric highs.
  C. Endorphins are addictive; athletes suffer withdrawal symptoms when they can't exercise.

18.10 No Smoking
   A. Smoking leads to health problems such as emphysema, chronic bronchitis, heart attacks, strokes, and cancer.
      1. Smokers are more likely to develop Alzheimer's disease.
   B. Second-hand smoke can be just as detrimental to nonsmokers as smoking is to smokers.

18.11 Chemistry of Sports Materials
   A. Athletic materials have been transformed by chemistry.
      1. Protective helmets and pads are products of the chemical industry.
      2. Playing fields have "floors" and roofs of synthetic materials.
      3. Table 18.3 in the text provides an extensive list of sport materials produced by chemists from petroleum.
   B. Goretex is a thin, membrous material with billions of tiny holes that are too small to allow drops of water or rain to pass through and wet the wearer, but large enough for perspiration in the form of water vapor to travel out, thus keeping the wearer dry.

CHAPTER OBJECTIVES

(You should...)

1. Be able to predict which type of muscles (fast-twitch or slow-twitch) are needed for a specific type of activity.

2. Know several characteristics of each type of muscle fiber.

3. Know that exercise increases the myoglobin content of skeletal muscle.

4. Know that the main nutritional need of an athlete as compared to a sedentary individual is more carbohydrates.

5. Know that athletes do not need extra protein in the diet to build muscles; only exercise builds muscles.

6. Know that muscle cannot "turn into fat."

7. Know that creatine, released when a muscle contracts, stimulates the production of myosin thus building more muscle.

8. Know that diets of fewer than 1200 Calories/day are not capable of providing sufficient nutrients.

9. Be able to apply the law of conservation of energy to weight loss or weight gain.

10. Know that 1 lb of adipose tissue contains 3500 Calories.

11. Know that quick-weight-loss diets depend on dehydration or glycogen depletion or both.

12. Know that any weight loss due to restricted carbohydrate intake results in a loss of muscle as well as fat.

13. Know that weight regained (without exercise) after dieting is pure fat.

14. Be able to describe carbohydrate loading.

15. Be able to describe blood doping.

16. Know that sodium ions and potassium ions are the two main electrolytes necessary for proper cellular function.

17. Know the end products of aerobic exercise.

18. Know the end products of anaerobic exercise.

19. Know that some energy from cellular metabolism is stored in ATP molecules.

20. Know that the immediate source of energy for muscle contraction is ATP, but the energy in ATP is sufficient for only a few seconds at best.

21. Know that when muscle contraction begins, glycogen is converted to pyruvic acid.

22. Know that the buildup of lactic acid lowers the pH of muscle tissue; this causes a weaker response of muscle cells to stimuli, known as muscle fatigue.

23. Know that relatively brief, intense exercise produces an oxygen debt; the athlete continues to gulp air after the event is over.

24. Know that fats are the main source of energy for exercise sustained over several hours.

25. Know that glycogen is the main source of energy for exercise for the first hour or so.

26. Know that the best liquid to replace fluids lost during exercise is plain water.

27. Know that alcohol and caffeine are diuretics; therefore, beer, coffee, and cola drinks are not good replacement fluids for water lost during exercise.

28. Know that thirst is not a good indication of dehydration.

29. Know that alcohol acts as a diuretic by blocking the action of antidiuretic hormone.

30. Know that sustained vigorous exercise causes the body to produce enkephalins and endorphins.

31. Know that restorative drugs are those used to alleviate the effects of exercise, such as soreness and slight injuries.

32. Know that anabolic steroids might have some effect in increasing muscle mass more rapidly, but they have many serious side effects.

33. Be able to describe how the development of new materials has affected several different sports.

34. Be able to tell what Goretex is and describe how it works.

35. Be able to calculate the amounts of exercise (distance or time) required to burn off a given amount of food Calories (kcal).

DISCUSSION

In this chapter, we consider application of chemistry to sports and athletics. Science has transformed many athletic events. We now have a better understanding of the action of muscles and of the effects of diet and exercise. Chemistry has provided a plethora of new materials for clothing and equipment. It also has provided drugs that help (and harm!) athletes. Chemical principles can help you to separate fact from fancy and to guard against quackery and harmful practices.

SELF-TEST

Multiple Choice

1. Exercise in which the body tissues have insufficient oxygen to oxidize glucose completely to carbon dioxide and water is called

    a. aerobic
    c. diuretic

    b. anaerobic
    d. doping

2. After engaging in a short, intensive burst of exercise, an athlete has

    a. dehydration
    c. more slow-twitch muscles

    b. electrolyte depletion
    d. an oxygen debt

3. In muscle tissue, anaerobic oxidation results in the production of

    a. carbon dioxide
    c. glycogen

    b. glucose
    d. lactic acid

4. Contraction of a muscle against a resistance causes production of a chemical that stimulates the production of myosin, a muscle protein. This chemical is

    a. creatine
    c. lactic acid

    b. glycogen
    d. pyruvic acid

5. When muscles are not used, they

    a. atrophy
    c. remain unchanged

    b. grow faster
    d. turn to fat

6. Tiredness and muscle pain are due to a buildup of

    a. fat
    c. lactic acid

    b. glycogen
    d. myoglobin

7. Some birds migrate nonstop for thousands of miles. Their flight fuel is mainly

    a. fats
    c. glycogen

    b. glucose
    d. proteins

8. The end product of aerobic oxidation in muscle tissue is

    a. amylose
    b. carbon dioxide
    c. glycogen
    d. lactic acid

9. Which activity would be aided by a high proportion of fast-twitch muscle?

    a. marathon run
    b. hard day's manual labor
    c. lift a 400-lb weight
    d. swim the English Channel (20 mi)

10. The only safe way to build lots of muscles is to

    a. eat lots of protein
    b. take lots of vitamins
    c. take anabolic steroids
    d. exercise

11. A runner is eating a balanced diet. If she wishes to increase her mileage, she should also increase her intake of

    a. carbohydrates
    b. fats
    c. proteins
    d. vitamins

12. On the first day of a fast, the body obtains energy mainly from

    a. fat
    b. blood glucose
    c. glycogen
    d. proteins

13. A dieter loses 7 lb in the first week on Doc Hill's "revolutionary diet plan." She has lost

    a. 7 lb of fat
    b. 5 lb of glycogen and 2 lb of fat
    c. 2 lb of glycogen and 5 lb of fat
    d. mostly water

14. Your favorite chemistry professor runs 10 miles on a warm day. He weighs 6 lb less at the end of the run than he did at the beginning. The weight loss is mainly

    a. fat
    b. glycogen
    c. muscle
    d. water

15. A prudent weight-loss diet will cause the dieter to lose not more than

    a. 1 lb/day
    b. 2 lb/day
    c. 2 lb/week
    d. 10 lb/week

16. The two main electrolytes necessary for proper cellular function are

    a. Ca^{2+} and Fe^{2+}
    b. Cl^- and SO_4^{2-}
    c. Na^+ and K^+
    d. ERG and Gatorade

17. Which is a good indicator of dehydration?

    a. a craving for salt
    b. a desire for a beer
    c. thirst
    d. none of these

18. To replace water lost during exercise, you should drink

a. gatorade

b. beer

c. plain water

d. mineral water

19. Antidiuretic hormone (ADH) signals the kidney to

a. excrete more Na^+

b. excrete more K^+

c. excrete more urea

d. conserve water

20. To lose weight safely, one should

a. eliminate carbohydrates from the diet

b. eliminate fats from the diet

c. eat a high-protein diet

d. eat a balanced diet with fewer Calories

21. You run 10 miles. Though tired, you feel just great because your brain has synthesized some

a. amphetamine

b. endorphins

c. morphine

d. antihistamines

22. Athletic performance can be improved by

a. anabolic steroids

b. electrolyte replacement fluids

c. high-protein diets

d. practice

23. Drugs used to relieve the soreness in muscles after an athletic event are called

a. anabolic steroids

b. antipyretics

c. restoratives

d. stimulants

24. Drugs that are thought to enhance the buildup of muscle tissue are called

a. anabolic steroids

b. analgesics

c. progestins

d. stimulants

25. Cocaine

a. increases strength

b. increases quickness

c. increases endurance

d. creates the delusion of invincibility

26. You can store about 500 g of glycogen with an energy content of 4 kcal/g. How far could you run on that glycogen if you expended about 100 kcal/km?

a. 2 km

b. 12.5 km

c. 20 km

d. 12,500 km

27. A Big Mac has an energy content of 1200 kcal. How far would you have to walk, at a rate of 5 km/hr, to use up that energy if you use up 300 kcal/hr?

a. 0.2 km

b. 2 km

c. 7 km

d. 20 km

28. A pound (0.45 kg) of fatty tissue contains about 3500 kcal of energy. How far would you have to run to burn off a pound of fat if your running burns off 100 kcal/km?

    a. 0.03 km   b. 0.35 km
    c. 3.5 km   d. 35 km

29. The percentage of calories from fat should be no more than

    a. 1%   b. 5%
    c. 10%   d. 25%

30. Athletes should get the extra calories they need from

    a. carbohydrates   b. fats
    c. proteins   d. vitamins

31. A lot of quick-weight-loss diets include diuretics. Diurectics

    a. double your use of calories   b. double your level of energy
    c. increase your output of urine   d. increase your use of vitamins

32. If your BMI is above 28%, you should

    a. increase your vitamins   b. increase your calorie intake
    c. decrease your water intake   d. lose weight

33. Vitamin C is

    a. ascorbic acid   b. water soluble
    c. an antioxidant   d. all of the above

34. Water soluble vitamins

    a. must be replaced regularly   b. are fat soluble
    c. are stored in the body   d. are red in color

35. Commercial sports drinks contain

    a. pure water   b. vitamins
    c. electrolytes   d. proteins

36. The greatest concern for smokers should be

    a. cancer   b. heart attacks and strokes
    c. weight gain   d. none of the above

37. Endurance activities use mostly

    a. slow-twitch muscles   b. fast-twitch muscles
    c. both slow- and fast-twitch muscles   d. none of the above

38. Oil of wintergreen is

        a. a stimulant   b. a depressant
        c. a hallucinogen   d. a counter irritant

39. Caffeine

        a. triggers the release of fatty acids   b. increases the heart rate
        c. speeds metabolism   d. all of the above

ANSWERS

1. b	8. b	15. c	22. d	28. d	34. a
2. d	9. c	16. c	23. c	29. d	35. c
3. d	10. d	17. d	24. a	30. a	36. b
4. a	11. a	18. c	25. d	31. c	37. a
5. a	12. c	19. d	26. c	32. d	38. d
6. c	13. d	20. d	27. d	33. d	39. d
7. a	14. d	21. b			

CHAPTER

19

Drugs

Chemical Cures, Comforts, and Cautions

KEY TERMS

Acquired Immune Deficiency Syndrome (AIDS)
agonists
allergen
analgesic
androgens
antagonists
antibiotics
anticoagulant
anti-inflammatory
antihistamine
antimetabolites
antipyretic
broad-spectrum antibiotics
chemotherapy

depressant drugs
dextro isomer
dissociative anesthetic
drug abuse
drug misuse
endorphins
estrogens
general anesthetic
hallucinogenic drug
hormones
levo isomer
local anesthetic
marijuana
narcotics

neurons
neurotransmitters
placebo
progestins
prostaglandins
psychotropic drug
retroviruses
steroids
stimulant drugs
steroids
synergistic effect
synapses

CHAPTER SUMMARY

19.1   Pain Relievers: Aspirin
    A. Aspirin (acetylsalicylic acid) was introduced in 1899 as one of the first successful synthetic pain relievers and has become the largest-selling drug in the world.
        1. Willow bark was a "home remedy" for reducing fever.
        2. Salicylic acid was isolated from willow bark in 1860. It was found to be a good <u>analgesic</u> (pain killer) and an <u>antipyretic</u> (fever reducer), but sour and irritating to take by mouth.
        3. Sodium salicylate (1875) and phenyl salicylate (1866) are chemical modifications of this natural drug (Figure 19.2 in the text), but both have undesirable attributes.
    B. Methyl salicylate (oil of wintergreen) is used as a flavoring and as a skin rub analgesic.
    C. Aspirin is the most widely used drug in the United States.
        1. Plain aspirin tablets have 325 mg each of acetylsalicylic acid (or 650 mg per two-tablet adult dose).
        2. Extra-strength formulations have 500 mg of acetylsalicylic acid per tablet.
           a.   Three regular tablets (975 mg) are virtually equivalent to two extra-strength tablets.
        3. Buffered aspirin contains small amounts of antacids; U.S. Food and Drug Administration (FDA) evaluations conclude that buffered aspirin is neither faster than plain aspirin nor easier on the stomach.

      a.   Taking aspirin after eating food or with a full glass of water is just as effective as buffered aspirin.

   D. Aspirin—pain relief and anti-inflammatory action
      1. Aspirin works by inhibiting the production of prostaglandins.
         a.   Prostaglandins are responsible for sending pain messages to the brain.
         b.   Inflammation results from over-production of prostaglandin derivatives.
         c.   The anti-inflammatory action of aspirin results from inhibition of prostaglandin synthesis.

   E. Aspirin as an anticoagulant—heart attack and stroke prevention
      1. Aspirin acts as an anticoagulant—it inhibits clotting of blood.
      2. Aspirin should not be used by those anticipating surgery or childbirth.
      3. Small doses of aspirin lower the risk of coronary heart attack and stroke.

   F. Aspirin and Fever Reduction
      1. Fevers are induced by pyrogens, which are produced by and released from leukocytes.
         a.   Pyrogens use prostaglandins as secondary mediators; therefore, fevers are reduced by inhibiting prostaglandins.
      2. Fevers are the body's mechanism for fighting infections.

   G. Limitations of Aspirin
      1. Not effective for severe pain.
      2. Can be toxic, especially to small children.
      3. Some people have an allergic reaction to it, including skin rashes, asthma attacks, and even loss of consciousness.

   H. Aspirin Substitutes: Acetaminophin and Ibuprofen
      1. Acetaminophen (Tylenol, Panadol, Anacin-3, generic)
         a.   Analgesic—reduces pain.
         b.   Antipyretic—reduces fevers.
         c.   Not anti-inflammatory (of little use to arthritis sufferers).
         d.   Does not promote bleeding (can be used by surgical patients).
         e.   Regular acetaminophen tablets are 325 mg; extra-strength tablets are 500 mg.
         f.   Overuse linked to liver and kidney damage.
      2. Ibuprofen (Motrin-IB, Advil, Nuprin, Medipren)
         a.   Analgesic—available in 200 mg tablets over the counter.
         b.   Anti-inflammatory—may be superior to aspirin in this function.
         c.   Antipyretic.

   I. Combination Pain Relievers
      1. The historical combination was a tablet that contained aspirin, phenacetin, and caffeine (APC).
         a.   Phenacetin is about as effective as aspirin as an analgesic and antipyretic.
         b.   Phenacetin has been implicated in kidney damage, blood abnormalities, and as a possible carcinogen.
         c.   Phenacetin was banned by the FDA in 1983.
      2. Anacin contains aspirin (400 mg) and caffeine.
         a.   Caffeine is a mild stimulant that has not been shown to be effective for pain relief at the dosages commonly used. As a stimulant, it probably is counterproductive in treating fevers.
      3. Excedrin contains aspirin, acetaminophen, and caffeine.
      4. Brands and formulations change frequently.
         a.   Plain aspirin is the cheapest, safest, and most effective product for most purposes.

19.2   Chemistry, Chicken Soup, and the Common Cold
   A. There is no cure for the common cold. Colds are caused by viruses.
      1. Cold remedies treat cold symptoms.
   B. Antihistamines and allergies
      1. Antihistamines (diphenhydramine, chlorpheniramine, and promethazine) relieve symptoms of allergies.

2. Allergies trigger the release of histamines which cause redness, swelling, and itching.
   a. Antihistamines block their release, but may cause drowsiness.
3. The prescription drugs (Seldane and Hismanal) inhibit release of histamine but cannot enter the brain and therefore do not cause drowsiness.
4. Antihistamines are not effective against colds.
C. Cough suppressants (antitussives) have three effective ingredients.
   1. Codeine, a narcotic.
   2. Dextromethorphan, a narcotic.
   3. Diphenhydramine, an antihistamine.
D. Expectorants and nasal decongestants
   1. Expectorants help bring up mucus from bronchial passages.
      a. Guaifenesin is the only safe and effective expectorant currently on the market.
   2. Nasal decongestants are safe for occasional use: phenylephrine, ephedrine, and their salts.
E. Chicken soup supplies liquid; the spices in it loosen nasal congestion; it's as good as anything for a cold.

19.3 Antibacterial Drugs
A. In 1900 infectious diseases were the principal cause of death.
   1. Antibacterial drugs have dramatically altered that situation.
B. Gerhard Domagk discovered sulfa drugs (the first antibacterial drugs) in 1935.
   1. Used extensively in WWII to prevent would infections.
C. Sulfa compounds inhibit the growth of bacteria by mimicking para-aminobenzoic acid (PABA), a nutrient needed by bacteria for proper growth.
   1. Bacteria mistake the sulfa drug (sulfanilamide) for PABA and produce molecules that cannot perform growth-enhancing functions.
      a. As a result, bacteria die.
D. Penicillins and Cephalosporins
   1. Penicillin, the first antibiotic (substances derived from molds or bacteria that inhibit the growth of other microorganisms), was discovered by Fleming in 1928.
      a. Florey and Chain purified penicillin for use in medicine.
   2. There are several penicillins that vary in structure and properties.
   3. Penicillin inhibits the synthesis of bacterial cell walls; without the walls, the cells collapse and die. (Human cells don't have cell walls.)
   4. Disadvantages of penicillin.
      a. Many people are allergic to it.
      b. Many kinds of bacteria have developed strains that are resistant to penicillins.
   5. Penicillins have been partially replaced by related compounds called cephalosporins.
      a. Keflex is an example of a cepholosporin.
      b. Some bacterial strains are resistant to cephalosporins.
E. Tetracyclines: Four Rings and Other Things
   1. The tetracycline antibiotics are characterized by four rings joined side to side.

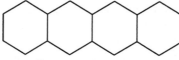

      a. Aureomycin (chlorotetracycline) was isolated in 1948.
      b. Terramycin was isolated in 1950.
      c. Tetracycline (the parent compound) was isolated in 1953.
   2. Tetracyclines are broad-spectrum antibiotics.
      a. They are effective against a wide variety of microorganisms.
      b. They bind to bacterial ribosomes.
      c. This inhibits bacterial protein synthesis and thus blocks bacterial growth.
   3. Disadvantages:
      a. Can cause discoloration of teeth in children.

19.4   Viruses and Antiviral Drugs
   A. Viral diseases (colds, flu, herpes, AIDS) are not amenable to treatment with antibiotics.
       1. Viral diseases are best dealt with by prevention. Vaccination prevents mumps, measles, and other dread viral diseases.
   B. DNA and RNA viruses
       1. Viruses are composed of nucleic acids and proteins.
           a. Genetic material may be DNA or RNA.
       2. DNA viruses replicate in host cells and direct production of viral proteins, which together with the viral DNA, assemble into new viruses.
           a. These viruses then invade other cells.
       3. RNA virus replication is similar to DNA.
           a. Some RNA viruses called retroviruses synthesize DNA in host cells.
               i. AIDS virus is a retrovirus that destroys T cells, which protect the body from infections.
   C. Antiviral drugs
       1. A few modestly effective antiviral drugs have been found.
           a. Amantadine has had some success in preventing influenza A infection.
           b. Acyclovir helps control (but does not cure) herpes infections.
           c. Azidothymidine (AZT) slows the onslaught of AIDS.
   D. Basic Research and Drug Development
       1. Gertrude Elion, George Hitchings, and James Black helped design antiviral drugs that block receptors in infected cells.

19.5   Chemicals Against Cancer
   A. Antimetabolites: Inhibition of nucleic acid synthesis
       1. Anticancer metabolites block DNA synthesis.
           a. Cancer cells that are dividing rapidly need large quantities of DNA and therefore are greatly affected by the DNA shortage.
               i.   Cisplatin binds to DNA and blocks its replication.
               ii. 5-Fluorouracil and 5-fluorodeoxyuridine inhibit the formation of a thymine-containing nucleotide required for DNA synthesis.
               iii. 6-Mercaptopurine substitutes foradenine and guanine and thus inhibits the synthesis of nucleotides incorporating adenine and guanine, slowing DNA synthesis.
   B   Methotrexate substitutes for folic acid in competition for an enzyme, but it lacks the growth-enhancing function of folic acid. Cell division is slowed, and cancer growth retarded.
       1. Agents: Turning War Gases on Cancer
           a. Alkylating agents are reactive compounds that transfer alkyl groups (methyl, ethyl, etc.) to biologically important compounds such as the bases in DNA.
           b. War gases
               i.   Mustard "gas" is a sulfur-containing blister agent used in chemical warfare in World War I.
               ii. Nitrogen mustards were developed about 1935 as chemical warfare agents.
               iii. Nitrogen mustards such as cyclophosphamide are effective anticancer drugs.
       2. Miscellaneous Anticancer Agents
           a. Alkaloids from vinca plants are effective against leukemia and Hodgkin's disease.
           b. Actinomycin from molds is used against Hodgkin's disease and other cancers.
               i.   It binds to the double helix of DNA, blocking the formation of RNA on the DNA template. Protein synthesis is inhibited.
           c. Sex hormones of the opposite gender can be used against breast cancer and prostate cancer.
           d. BHT, vitamin A, and vitamin C may have some anticancer activity.

19.6   Hormones: The Regulators
   A. Hormones are chemical messengers produced in the endocrine glands.

1. They cause profound changes in parts of the body often far removed from the gland that secretes the hormone.
2. Table 19.2 in the text lists a variety of hormones and their physiological effects.

B. Prostaglandins: Hormone Mediators and Medicines
1. Prostaglandins are synthesized from unsaturated fatty acids that contain 20 carbon atoms (such as arachondic acid).
2. Prostaglandins act together with hormones to regulate smooth-muscle activity and blood flow.
3. Medically, prostaglandins and derivatives are used to
   a. Induce labor.
   b. Lower or raise blood pressure.
   c. Inhibit stomach secretions.
   d. Relieve nasal congestion.
   e. Relieve asthma.
   f. Inhibit the formation of blood clots.

C. The Steroids
1. All steroids have the same skeletal four-ring structure.
2. Not all steroids have hormonal activity.
   a. Cholesterol is a steroid but not a hormone.
   b. Cortisone is a steroid hormone produced by the adrenal glands.
      i. Medically, it and related compounds such as prednisone are anti-inflammatory drugs.
3. Steroid Drugs
   a. Anabolic steroids are taken to improve athletic performance.
      i. Also used to treat arthritis and bronchial asthma.
4. The Sex Hormones
   a. Testosterone, produced in the testes, is the principal male hormone.
   b. Androgens are responsible for development of the sex organs and for secondary sexual characteristics such as voice and hair distribution.
   c. Two important groups of female hormones are the estrogens and progesterone.
      i. Estrogens, produced in the ovaries, are female hormones.
      ii. Two important estrogens are estradiol and estrone.
      iii. Estrogens control the menstrual cycle and are responsible for secondary sexual characteristics such as the growth of breasts.
      iv. Progesterone is a female hormone that prepares the uterus for pregnancy and prevents the further release of eggs from the ovaries.
4. Sex hormones are used medically to treat certain types of cancer.

19.7   Chemistry and Social Revolution: The Pill
A. Progesterone is an effective birth control drug when injected.
B. Synthetic analogs of progesterone, called progestins, are effective birth control drugs. Oral progestins incorporate an ethynyl group.
C. Oral birth control pills usually combine an estrogen (to regulate the menstrual cycle) with a progestin that signals a state of false pregnancy so that ovulation does not occur.
D. DES: A Missed-Period Pill
1. Diethylstilbestrol (DES) has had three uses.
   a. In small doses to prevent miscarriages.
      i. Shown to cause cancer in daughters of women who took DES for this purpose.
   b. In large doses to induce abortion.
   c. In animal feed to help induce weight gain in animals.
      i. Banned by the FDA when DES residues were found in meat from these animals.
2. Under the Delaney Amendment, DES is banned from appearing in food because it may lead to cancer, but this amendment does not regulate drugs.
E. RU-486: Convenience and controversy

1. Mifepristone (RU-486) blocks the action of progesterone; as a result a pregnancy is terminated.
    a. RU-486 can increase uterine contractions, trigger lactation in mothers, and slow the growth of some cancers.
    b. RU-486 is available in France and China.
 F. Risks of taking the pill
    1. Side effects include blood clotting (which can cause stroke and heart attack) in some women.
    2. Overall, the risk of taking the pill is one-tenth that of having a baby.
 G. Mini-pill—-small amount of progestin and no estrogen
    1. It avoids the problems of side effects, which are due to the presence of estrogen.
 H. Birth Control by Implant
    1. Norplant, a synthetic progesterone analog, releases progesterone into the bloodstream over time.
 I. A Pill for Males?
    1. Women usually bear the responsibility for contraception because
       a. They get pregnant if contraception fails.
       b. It is easier to interfere with a once-a-month event (ovulation) than a continuous process (sperm production).
       c. Gossypol, a pigment from cottonseed, has been used in China as a contraceptive for males.
       d. Availability of vasectomies has reduced the demand for a male contraceptive.

19.8 Drugs for the Heart
 A. Cardiovascular drugs rank first in worldwide sales of prescription drugs.
 B. Hypertension can be relieved by diuretics (lowers the volume of blood), beta blockers (slow the heart rate), calcium channel blockers (induce muscles around blood vessels to relax), and ACE (angiotension converting enzyme)
 C. Arrhythmia (abnormal heartbeat) can be treated with a variety of drugs with varying mechanisms of action.
 D. Angina pectoris (insufficient oxygen to the heart) is often treated by dilating the blood vessels around the heart.

19.9. Drugs and the Human Mind
 A. Psychotropic drugs affect the mind. There are three types:
    1. <u>Stimulants</u> increase alertness and elevate the mood.
       a. Amphetamines, caffeine, and cocaine are stimulants.
    2. <u>Depressants</u> reduce the level of consciousness and the intensity of reactions to outside stimuli.
       a. Ethanol, barbiturates, opiates, and tranquilizers are depressants.
    3. <u>Hallucinogenic</u> (psychotomimetic, psychedelic) drugs alter qualitatively the way we perceive things.
       a. LSD mescaline, and marijuana are hallucinogens.

19.10 Some Chemistry of the Nervous System
 A. Nerve cells (neurons) carry messages between the brain and other parts of the body.
    1. <u>Axons</u> of a nerve cell can be very long; however, the nerve impulse must be transmitted to the next dendrite across short fluid-filled gaps (synapses) via chemical messengers called neurotransmitters.
       a. Each type of neurotransmitter binds to a specific type of receptor site to complete the intended action.
 B. Many drugs (and poisons) act either by blocking or mimicking the action of these natural neurotransmitters.
 C. Many neurotransmitters are amines.

19.11 Brain Amines: Depression and Mania
 A. Epinephrine (adrenaline) is an amine produced by the adrenal glands.
    1. Adrenaline prepares the body for <u>fight</u> or <u>flight</u>.
 B. A Biochemical Theory of Mental Illness

1. The neurotransmitter norepinephrine is chemically related to the hormone adrenaline (epinephrine).
    a. Norepinephrine is synthesized from the amino acid tyrosine.
    b. High levels of norepinephrine are associated with a state of elation.
    c. Drugs that block the action of norepinephrine are depressants; drugs that mimic its action are stimulants.
C. Other Brain Amines
    1. Serotonin, a neurotransmitter, is involved in sleep, sensory perception, and regulation of body temperature.
       a. Low levels of a serotonin metabolite have been found in suicide victims.
D. Brain Amine Agonists and Antagonists in Medicine
    1. Norepinephrine (NE) agonists (drugs that enhance or mimic its action) are stimulants.
    2. NE antagonists (drugs that block its action) slow down processes.
       a. Beta blockers are an example.
    3. Serotonin agonists are used to treat depression.
    4. Serotonin antagonists are used to treat obsessive-compulsive behavior.

E. Brain Amines and Diet: You Feel What You Eat
    1. Serotonin is synthesized from the amino acid tryprophan.
       a. A high-carbohydrate meal allows maximum tryptophan to reach the brain (where it is converted to serotonin).
       b. Protein-rich diets lower the level of serotonin in the brain.
    2. Norepinephrine is synthesized from the amino acid tyrosine.
F. Love: A Chemical Connection
    1. Phenylethylamine (PEA) is a neurotransmitter.
    2. Increased levels of PEA in the brain produce a "high" identical to that described as "being in love."
       a. Protein-rich foods contain the amino acid phenylalanine, which serves as a precursor to PEA.
    3. Low levels of the PEA metabolite phenylacetic acid in the urine correlate with depression.

19.12 Anesthetics
    A. General anesthetics render one unconscious and insensitive to pain.
       1. They are the ultimate depressant.
    B. History of anesthesia
       1. Diethyl ether (ether) was the first general anesthetic (1846).
          a. Ether is relatively safe, but causes nausea.
          b. Ether is highly flammable.
       2. Nitrous oxide is a quick-acting anesthetic. It is administered with oxygen to prevent brain damage.
       3. Chloroform was once used as an anesthetic.
          a. It has a narrow safety margin; the effective dose is close to the lethal dose.
          b. Causes liver damage.
          c. Can react with oxygen to form deadly phosgene gas.
          d. It is nonflammable.
    C. Modern anesthetics include fluorine-containing compounds such as halothane, enflurane, and isoflurane.
       1. These compounds are nonflammable and relatively safe for the patient.
          a. Women who work in operating rooms where halothane is used have higher rates of miscarriage than the general population.
       2. Modern surgical practice often combines the following:
          a. An intravenous anesthetic such as thiopental.
          b. An inhalant anesthetic.
          c. A muscle relaxant such as curare.
       3. The potency of an anesthetic is related to its solubility in fat. It acts by dissolving in and changing the permeability of cell membranes.
    D. Solvent Sniffing: Self-Administered Anesthesia

1. Organic solvents generally act as anesthetics; glue sniffing results in self-anesthesia.
  E. Local Anesthetics
    1. Local anesthetics render one part of the body insensitive to pain but leave the patient conscious.
      a. Cocaine was the first local anesthetic.
    2. Many local anesthetics are ester derivatives of, or related to, para-aminobenzoic acid.
      a. Procaine (Novocaine) was introduced in 1905.
      b. Lidocaine and mepivicaine are widely used today.
  F. Dissociative Anesthetics: Ketamine and PCP
    1. Ketamine is an anesthetic that also produces "near death" type hallucinations.
    2. PCP (phencyclidine), also known as "angel dust," is a dangerous drug, but has found use as an animal tranquilizer.
      a. PCP is fat soluble. Stored in body fat, it is mobilized when fat is metabolized, causing "flashbacks."
      b. PCP depresses the immune system.
      c. One in 1,000 users develops severe schizophrenia.

19.13   Depressants
  A. Ethyl alcohol has been used by man for centuries.
  B. It is a depressant.
    1. It slows down physical and mental activity.
  C. Alcoholism is the fifth leading cause of deaths in the United States.
  D. Alcohol is a potent teratogen (cause of birth defects).
  E. The Barbiturates: Sedation, Sleep, and Synergism.
    1. Barbituric acid was synthesized from urea and malonic acid in 1864 by von Baeyer.
      a. Barbiturates as medicine
        i.   In small doses barbiturates act as sedatives.
        ii. Pentobarbital (Nembutal) is a short-acting barbiturate used to calm anxiety.
        iii. Phenobarbital (Luminal) is a long-acting barbiturate used as a sedative and anticonvulsant.
      b. Large doses of barbiturates serve as sleeping pills.
    2. Synergism—barbiturates and alcohol
      a. The synergistic action of these two depressants can enhance the effect of the barbiturate by up to 200-fold.
      b. Barbiturates are strongly addictive, and withdrawal is hazardous.
    3. Barbiturates are cyclic amides that resemble thymine.
      a. They may act by substituting for pyrimidine bases in nucleic acids.

19.14   The Opium Alkaloids: Narcotics
  A. Opium and Morphine
    1. Opium is the dried, resinous juice of seeds of the Oriental poppy.
      a. It is a mixture of 20 alkaloids, sugars, resins, and waxes.
    2. Morphine is the principal alkaloid of opium (10% by weight).
  B. Morphine
    1. Morphine is a narcotic, producing sedation and analgesia.
    2. Morphine is strongly addictive.
    3. Morphine and other narcotics came under government control with the Harrison Act of 1914.
  C. Codeine and Heroin
    1. Codeine is methylmorphine.
      a. Codeine is less potent, less likely to induce sleep, and less addictive than morphine.
    2. Heroin is the trade name of diacetylmorphine.
      a. It was introduced by the Bayer Company of Germany in 1874 as a synthetic analgesic and antidote for morphine addiction.
      b. Heroin produces a strong feeling of euphoria.

     c.   Heroine produces addiction and is illegal in the United States.
  D. Synthetic Narcotics: Analgesia and Addiction
     1. Meperidine (Demerol) is a synthetic narcotic somewhat less potent than morphine.
     2. Methadone is used to treat heroin addiction. Methadone is addictive, but when taken orally it does not induce the sleepy stupor of heroin.
     3. More morphine analogs: agonists and antagonists.
        a. Thousands of morphine analogs have been synthesized.
        b. <u>Agonists</u> are substances that mimic drug action.
        c. <u>Antagonists</u> are substances that block the action of a drug.
        d. Some molecules have both agonist and antagonist action.
           i. Example: Pentazocine is less addictive than morphine but effective for relief of pain.
  E. A Natural High: The Brain's Own Opiates
     1. Morphine fits receptors in the brain.
     2. Several natural substances that fit those receptors are produced by the body. These substances have morphine-like activity.
        a. Enkephalins have five amino acid units; Leu-enkephalin and Met-enkephalin differ from each other by the one amino acid unit indicated in their names.
        b. Endorphins have chains of 30 amino acid units.
     3. Endorphins are released in response to pain deep in the body and are involved in the following phenomena:
        a. A soldier doesn't feel his wounds until the battle is over.
        b. The anesthetizing effect of acupuncture.

19.15 Antianxiety Agents
  A. Ethanol is the most widely used tranquilizer.
  B. Products such as Cope, Vanquish, and Compoze contain aspirin plus an antihistamine.
  C. Minor tranquilizers (benzodiazepine derivatives) are used to treat anxiety.
     1. The benzodiazepines contain a seven-membered heterocyclic ring system like the one in diazepam (Valium) and chlorodiazeposide (Librium).
        a. Certain benzodiazepines are used to treat insomnia (example: Halcion).
     2. Benzodiazepines act by fitting specific receptors. No natural Valium-like compounds have yet been found in our bodies. Rather, scientists have found terror-inducing compounds called ß-carbolines.
  D. Antipsychotic Agents
     1. Reserpine is the active alkaloid in the snake-root used by the people of India to treat fever, snake bite, and maniacal forms of mental illness.
        a. Reserpine reduces blood pressure and brings about sedation.
        b. By 1953, reserpine had replaced electroshock therapy for many psychotic patients as a leading antipsychotic agent.
     2. Penothiazines
        a. Promazine, chlorpromazine, and thioridazine (Mellaril) are effective against the symptoms of schizophrenia.
        b. Penothiazines are dopamine antagonists.
           i. Scientists believe schizophrenic patients produce too much dopamine or have too many dopamine receptors.
  E. Antidepressants
     1. Imipramine (Trofanil), in which a CH_2CH_2 group replaces the sulfur atom of promazine, is an antidepressant.
        a. Low doses of these tricyclic antidepressants have little effect. High doses can be toxic.
     2. New antidepressants, including Prozac, are safer.
        a. Prozac enhances the effect of serotonin by blocking its reabsorption by cells.

19.16   Stimulant Drugs: Amphetamines
A. The amphetamines form a group of synthetic stimulants, which may act by mimicking the natural brain amines epinephrine and norepinephrine.
   1.   All amphetamines are derived from phenylethylamine.
B. Amphetamine and methamphetamine (speed) are widely abused stimulant drugs.
C. Amphetamine exists as a pair of isomers that are mirror images of one another.
   1.   Dextroamphetamine, known as Dexedrine, (the "right-handed" isomer) is more potent than the levo ("left-handed") form.
D. Phenylpropanolamine is an over-the-counter appetite suppressant.
   1.   Widely used as a prescription "diet pill" for many years.
E. Methylphenidate (Ritalin) is used to treat hyperactivity in children.
F. Cocaine: The Snow Sniffers
   1.   Cocaine was first isolated from the leaves of the cocoa plant.
      a.   Cocaine arrives illegally in this country as broken lumps of the free base (crack cocaine).
      b.   Cocaine was once used as a local anesthetic, but it is quite toxic.
   2.   Cocaine acts by preventing the neurotransmitter dopamine from being taken back up from the synapse after it is released by nerve cells.
      a.   The cells fire wildly depleting the dopamine supply.
   3.   Cocaine is a stimulant, which increases stamina and reduces fatigue.
      a.   The stimulant effects are short-lived, followed by depression.
      b.   An overdose may cause death.
G. Caffeine: Coffee, Tea, or Cola
   1.   Caffeine is the most widely used stimulant.
      a.   Caffeine is an alkaloid found in coffee, tea, and some soft drinks.
      b.   Caffeine is available in No-Doz and Vivarin tablets.
   2.   The "morning grouch" syndrome and headaches upon withdrawal indicate that caffeine is addictive.
H. Nicotine: Going Up in Smoke
   1.   Nicotine is an alkaloid found in smoking and chewing tobacco.
      a.   The lethal dose (when injected) for humans is estimated at about 50 mg.
      b.   Nicotine is believed to be addictive.

19.17   The "Mind Benders": LSD
A. LSD (N, N-diethylamide of lysergic acid) was discovered by Hofmann in 1943.
   1.   It is a hallucinogen related to lysergic acid and other ergot fungus alkaloids.
   2.   LSD's structure resembles that of serotonin.
   3.   As little as 10 mg of this powerful drug can cause hallucinations.

19.18   Marijuana: Some Chemistry of Cannabis
A. Marijuana Forms and Potency
   1.   Marijuana is the leaves, flowers, seeds, and small stems of the <u>Cannabis</u> <u>sativa</u> plant.
B. The principal active ingredient is tetrahydrocannabinol (THC).
   1.   Typical smoking marijuana is about 1% THC; the potency depends on the genetic variety of the plant.
   2.   Hashish has a THC content of 5 to 12%.
C. Effects of Marijuana
   1.   Smoking marijuana increases the pulse rate, distorts the sense of time, and impairs complex motor functions. Many other effects have been reported.
   2.   Long-term effects are largely unknown. Heavy use has led to growth of breasts on males;
      a.   THC bonds to estrogen receptors.
      b.   THC produces an initial rise in testosterone in men. With high doses, this is followed by a rapid fall to below-normal testosterone levels.
D. Chemistry and marijuana

1. THC persists in the blood for days.
2. THC reduces pressure in eyes of glaucoma patients and relieves nausea of cancer patients undergoing radiation and chemotherapy.

19.19 Drugs and Deception: Chemistry and Quality Control
    A. On the illegal drug scene, drugs are not always what they are supposed to be.
        1. Potency varies widely; such variation can lead to death from accidental overdose.
            a. PCP, a cheap drug, is often sold as something more glamorous, such as "mescaline" or "synthetic THC."
    B. Drug misuse: inappropriate use of a drug to treat a specific illness.
    C. Drug abuse: using a drug for its intoxicating effect.

19.20 Placebo Effect
    A. Inactive substance given in the form of treatment of patient who thinks it is the real thing
        1. Patients receiving placebos often report positive effects.
        2. Placebo effect emphasizes the connection between mind and body.

CHAPTER OBJECTIVES

(You should know that...)

1. While chemicals have been used since the beginnings of history to treat ailments, modern "chemotherapy" began with Paul Ehrlich who realized that certain chemicals were more toxic to disease organisms than to human cells.

2. The most popular drug today is aspirin, a pain killer, anticoagulant, and fever reducer. Acetaminophen and ibuprofen are aspirin substitutes for common pain relievers as well as a combination mixture containing aspirin, phenacetin and caffeine. Each has its advantages and disadvantages.

3. Colds are caused by over 100 related viruses. No drug is effective but the symptoms can be alleviated.

4. Antihistamines relieve the symptoms of allergies (sneezing, itchy eyes, and runny nose) but are not effective for colds. Antihistamines work by blocking the release of histamine (which the body uses to repel the allergen) but antihistamines also tend to enter the brain and act on the cells controlling sleep.

5. Three compounds have been found effective as cough suppressants: codeine and dextromethorphan (which are narcotics) and an antihistamine, diphenylhydramine. The cough mechanism should be suppressed only to allow sleep as it is needed to remove congestion.

6. The FDA has found only guaifensin to be safe and effective as an expectorant, a chemical to bring up mucus from the bronchial passages but there are a number nasal decongestants such as phenylephrine and ephedrine safe and effective for occasional use.

7. Infectious diseases were the principal cause of death less than a century ago. Today only the infectious disease class, pneumonia and influenza, remains in the top ten causes of death. The first antibacterial drugs were the sulfa drugs but they are no longer used. AIDS is a new viral infection.

8. Penicillin, an antibiotic, replaced the sulfa drugs. These were followed by the tetracyclines.

9. Viruses are composed of nucleic acids and proteins.

10. Anticancer drugs are antimetabolites (inhibit the synthesis of nucleic acids) or alkylating agents (block the action of the molecules of cancer) and a number of a variety of anticancer agents with a wide range of properties.

11. Hormones are chemical messengers produced in the endocrine gland. They include prostaglandins (hormone mediators), which control a wide variety of physiological effects.

12. Steroids are hormones based upon a four-membered fused ring system. They include the sex hormones (and the contraceptive, "the pill").

13. Psychotropic drugs affect the human mind. Stimulants (cocaine and amphetamines) increase alertness, speed up the mind, and elevate the sprit; depressants (alcohols, most anesthetics, barbiturates, and opiates) reduce the responses to stimuli; and hallucinogens (LSD and marijuana) alter the perception.

14. Alcohol, the most widely used drug, was known to the ancients. It has serious health side effects including accidents, high suicide rates, and it is a potent teratogen (causes birth defects).

15. A general anesthetic acts on the brain to produce unconsciousness and a general insensitivity to feeling or pain. Diethyl ether was the first, followed by nitrous oxide and chloroform. Modern anesthetics include fluorine containing compounds such as halothen, enfluane, and methyloxyflurane. Most hydrocarbons that are associated with "sniffing" are anesthetics and are extremely deadly as their lethal dosages are close to their "intoxication" dosage.

16. Local anesthetics render one part of the body insensitive to pain while the patient remains conscious. They work by blocking nerve impulses to the brain.

17. Barbiturates, based upon barbituric acid, cause mild sedation, deep sleep, and even death. They are especially dangerous when taken with alcohol. Medical uses include hypnotic inducers and sedatives.

18. Narcotics are drugs that produce stupor or general anesthetics and relief of pain. The legal definition includes addictiveness. The opiates (morphine, codeine, and heroin) are related structurally.

19. Addiction has three components: emotional dependence, physical dependence, and tolerance. Psychological dependence is evident in the uncontrollable desire for a drug. Physical dependence is evident by acute withdrawal symptoms such as convulsions. Tolerance is a need for increasing dosages to produce the same effect.

20. There are several synthetic narcotics including meperidine (Demerol) and methadone. Chemists have also made morphine agonists (drugs that copy the actions of morphine) and antagonists (drugs that block the action of morphine by blocking receptor sites).

21. Endorphins are natural pain relievers produced in the brain. Acupuncture and the "runner's high" are explained by endorphins.

22. The nervous system consists of neurons (nerve cells) and synapses (gaps between the nerve cells). Chemicals called neurotransmitters carry the electrical signals between nerve cells.

23. An excess of norepinephrine, a neurotransmitter, causes elation; a deficit of norepinephrine causes depression. Norepinephrine agonists are stimulants while norepomephrine antagonists are depressants. Drugs and diet can effect the levels these chemicals.

248

24. Antianxiety drugs work by dulling and making insensitive stimuli. They include alcohol, valium, an antipsychotic drug called reserpine, phenothiazines used as tranquilizers and antidepressants such as Prozac.

25. Amphetamines are stimulants that include Ritalin, the cocaine family, caffeine, and nicotine.

26. LSD is one of the most powerful "mindbenders." It seems to work as a serotonin agonist.

27. Marijuana is the second most widely used intoxicant drug. The active ingredient is THC, which increases the pulse rate, distorts the passage of time, and impairs some complex motor functions. While there is little evidence that marijuana use leads to "harder" drugs, there is some evidence for brain damage.

28. Illegal drugs are dangerous because they are not always what they are said to be. Drug misuse, for example, using penicillin to cure a cold (which it can't) and drug abuse (using a drug for its intoxicating effect) is a widespread problem (more money is spent on illegal drugs than on food worldwide).

29. A placebo is an inactive substance given in the form of medication to a patient who thinks it is the real thing. That many people report positive results from taking placebos proves a strong connection between the mind and the body.

DISCUSSION

The first half of this chapter concentrates on chemical substances that are used to relieve pain or distress, to cure or alleviate disease, to prevent pregnancy, and for a variety of other purposes. Once again, keep in mind that a chemical substance has a specific set of properties that are invariant. Each compound may have some properties that are desirable and some that are undesirable; drugs may have nasty side effects as well as desired therapeutic properties.

The second half of this chapter is devoted to drugs that affect our mental state. It is important to realize that our moods, our sense of feeling "up" or "down," and our sense of and tolerance of pain are all influenced by the presence or absence of chemical molecules that help regulate our body functions. We now know that many of the drugs that have been used and abused for centuries are capable of eliciting certain effects because they somehow mimic a natural neurotransmitter or body regulator molecule. This usually involves some structural similarity that allows the drug molecule to bind to a receptor site intended for the normal body regulator molecule.

SELF-TEST

Multiple Choice

1. Aspirin is a

    a. single chemical compound
    b. mixture of acetylsalicylic acid and caffeine
    c. mixture of a variety of chemical substances
    d. drug derived from plant sources

2. Aspirin is

    a. salicylic acid
    c. methyl salicylate
    b. acetylsalicylic acid
    d. phenyl salicylate

3. The Phanciphul Pharmaceutical Company claims that its over-the-counter product contains more pain reliever than a competitive brand. Regardless of the claim, the pain reliever is most likely

    a. aspirin
    c. codeine
    b. caffeine
    d. Darvon

4. Arthritis Pain Formula contains 500 mg of aspirin per tablet. To get nearly the same amount of aspirin as in two APF tablets, you could take

    a. two 325-mg aspirin
    c. two Extra-Strength Tylenol tablets
    b. three 325-mg aspirin tablets
    d. two Anacin tablets

5. The "pain reliever doctors recommend most" is

    a. Tylenol
    c. aspirin
    b. Darvon
    d. caffeine

6. The APC combination refers to which of the following?

    a. aspirin, phenol, and cocaine
    b. alcohol, prune juice, and carrot juice
    c. aspirin, phosphates, and Cope
    d. aspirin, phenacetin, and caffeine

7. Which of the following compounds has been used extensively as a substitute drug for those who are allergic to aspirin?

    a. acetaminophen
    c. thalidomide
    b. LSD
    d. morphine

8. An advantage of acetaminophen, the aspirin substitute, over aspirin is that it

    a. is better at relieving pain
    c. is better at reducing inflammation
    b. is better at reducing fever
    d. may be taken by people who are allergic to aspirin

9. Which of the following has been clinically proven to be more effective than plain aspirin for the relief of headache pain?

    a. Excedrin   b. Anacin
    c. Vanquish   d. none of these

10. Anacin is

    a. a chemical compound similar to aspirin
    b. a mixture of aspirin and caffeine
    c. more effective than aspirin
    d. all of these

11. For relief of pain associated with a tooth extraction, a dentist often recommends acetaminophen because

    a. it is more effective than aspirin   b. it reduces inflammation
    c. aspirin promotes bleeding   d. it is always safer than aspirin

12. Excedrin contains aspirin and an aspirin substitute. The substitute has what advantage over plain aspirin for relief of headache pain?

    a. extra strength   b. lower toxicity
    c. faster action   d. none of these

13. Which medication shortens the duration of a common cold?

    a. penicillin   b. Contac
    c. Nyquil   d. none of these

14. Substances that trigger the release of histamines are called

    a. allergens   b. antihistamines
    c. antimetabolites   d. viruses

15. A drug that is effective as an antibacterial agent because it has the same general molecular shape as para-aminobenzoic acid (PABA) is

    a. penicillin   b. aureomycin
    c. sulfanilamide   d. terramycin

16. Which antibiotic kills bacteria by interfering with the synthesis of bacterial cell walls?

    a. sulfa drugs   b. penicillins
    c. tetracyclines   d. prostaglandins

17. Which antibacterial agent retards bacterial growth by interfering with bacterial protein synthesis?
    a. penicillin   b. sulfanilamide
    c. tetracycline   d. prostaglandin

18. Which drug is somewhat effective against viral infections?

    a. acyclovir   b. cephalexin
    c. penicillin   d. tetracycline

19. Chemical messengers produced by the endocrine glands are called

    a. histamines   b. hormones
    c. neurotransmitters   d. prostaglandins

20. At present, the best way to deal with most viral diseases is

    a. penicillin injections   b. sulfanilamide drugs
    c. tetracycline pills   d. vaccination

21. Which substance is a steroid, but not a hormone?

    a. cholesterol   b. cortisone
    c. progesterone   d. testosterone

22. Male sex hormones are called

    a. androgens   b. estrogens
    c. progestins   d. prostaglandins

23. The hormone that prepares the uterus for pregnancy and prevents the further release of ova is

    a. testosterone   b. estradiol
    c. progesterone   d. ethisterone

24. The most common birth control pill contains

    a. an estrogen and a progestin   b. an androgen and a progestin
    c. an estrogen and an androgen   d. diethylstilbestrol (DES)

25. Which anticancer drug is an antimetabolite?

    a. 5-fluoroaracil   b. cyclophosphamide
    c. actinomycin   d. all of these

26. Which of the following has been used as a successful anticancer drug?

    a. 5-fluoroaracil   b. LSD
    c. aminotriazole   d. sulfanilamide

27. Which of these is a common anesthetic in use today?

    a. dichloromethane   b. halothane
    c. phencyclidine   d. rauwolfia

28. Which of the following is a barbiturate?

    a. codeine   b. mescaline
    c. promethazine   d. thiopental

29. The interaction of two drugs to give an effect markedly greater than that of either alone is called

    a. addiction   b. addition
    c. synergism   d. stimulation

30. Which of the following is not chemically related to (derived from) morphine?

    a. heroin   b. codeine
    c. cocaine   d. none are related

31. Which drug is not a stimulant?

    a. caffeine   b. amphetamine
    c. cocaine   d. ethyl alcohol

32. Which of these has been shown not to be addictive?

    a. caffeine   b. nicotine
    c. ethyl alcohol   d. none

33. Which substance would make you "drunk"?

    a. amphetamine   b. pentobarbital
    c. caffeine   d. cocaine

34. Heroin is a chemical derivative of

    a. morphine   b. caffeine
    c. aspirin   d. barbituric acid

35. Natural painkillers, formed in the brain, are released in case of severe injury or vigorous physical activity. These compounds give us natural highs. They are called

    a. alcohol   b. amines
    c. enkephalins   d. alkaloids

36. A combination of alcohol and a barbiturate has an effect up to 200 times that of either alone. Such an effect is called

    a. synergism   b. chemical dependency
    c. synthesis   d. psychological addiction

37. The brain amine associated with natural highs is

    a. norepinephrine   b. methamphetamine
    c. mescaline   d. methylphenidate

38. Esters of para-aminobenzoic acid are used widely as

   a. "uppers"
   c. hallucinogens

   b. "downers"
   d. local anesthetics

39. Reserpine and the promazines are used

   a. in the treatment of mental illnesses
   c. as general anesthetics

   b. as local anesthetics
   d. as sleeping pills

40. Which class of compounds is related to phenylethylamine?

   a. alkaloids
   c. barbiturates

   b. amphetamines
   d. tranquilizers

41. Chemically, amphetamine is an

   a. acid
   c. alcohol

   b. aldehyde
   d. amine

42. PCP, known on the illegal drug market as "angel dust" is a(n)

   a. amphetamine
   c. animal tranquilizer

   b. barbiturate
   d. plant alkaloid

43. Which of the following is a dissociative anesthetic?

   a. cocaine
   c. ketamine

   b. halothane
   d. thiopental

44. Drugs that cause changes in visual perception, depersonalization, and other brain disturbances are classed as

   a. narcotics
   c. downers

   b. uppers
   d. hallucinogens

45. The hallucinogenic drug LSD is a derivative of which chemical obtained from the ergot fungus, which grows on rye?

   a. ergotine
   c. lysergic acid

   b. mescaline
   d. diethylamine

46. The variation in the tetrahydrocannabinol (THC) content of marijuana depends mainly on the

   a. geographic area where the plant is grown
   b. climate in which the plant is grown
   c. genetic variety of the plant
   d. soil in which the plant is grown

47. Which of the following drugs remains in the bloodstream the longest?

   a. aspirin
   c. alcohol

   b. caffeine
   d. tetrahydrocannabinol

48. Marijuana has been clinically shown to

    a. be perfectly safe
    b. result in no physiological change in its users
    c. result in no birth defects in the offspring of users
    d. none of these

49. Your drug supplier offers you a "new" marijuana called "Wisconsin Weed." The material is much more potent than the plant that grows wild in the northern United States. The weed is most likely to be

    a. a new genetic variety
    b. nonpotent marijuana laced with PCP
    c. Mexican marijuana
    d. catnip

50. Clinical studies of marijuana have been hampered by a lack of

    a. reliable subjects   b. placebos
    c. marijuana of standard potency   d. trained investigators

51. A "friend" offers you some "mescaline" for $5.00. Chances are that the stuff is really

    a. mescaline   b. tetrahydrocannabinol (THC)
    c. phencyclidine (PCP)   d. lysergic acid diethylamide (LSD)

52. The Putrid Palliative Pill Company introduces a new over-the-counter stimulant drug called "Elevator." The most likely active ingredient is

    a. nicotine   b. caffeine
    c. amphetamine   d. cocaine

53. Gourmet Goofballs, Inc., introduces a new prescription drug for use as a sleeping pill. Some of the drug is diverted to the illegal drug market, where it is used as an intoxicant. The drug is most likely a(n)

    a. amphetamine   b. hallucinogen
    c. barbiturate   d. antihistamine

54. Brain Busters, Inc., comes out with a new drug that alters visual perception but has neither depressant nor stimulant activity. The drug is a(n)

    a. amphetamine   b. barbiturate
    c. antihistamine   d. hallucinogen

55. Your friendly neighborhood "supplier" offers you some "synthetic tetrahydrocannabinol (THC)." Analysis in a crime laboratory would most likely show that the drug is really

    a. genuine   b. PCP
    c. cocaine   d. aspirin

56. Which is a completely synthetic drug?

        a.   tetrahydrocannabinol (THC)   b.   morphine
        c.   cocaine   d.   amphetamine

57. Which of these is a single chemical compound?

        a.   opium   b.   heroin
        c.   peyote   d.   marijuana

ANSWERS

1. a	11. c	21. a	31. d	41. d	51. c
2. b	12. d	22. a	32. d	42. c	52. b
3. a	13. d	23. c	33. b	43. c	53. c
4. b	14. a	24. a	34. a	44. d	54. d
5. c	15. c	25. a	35. c	45. c	55. b
6. d	16. b	26. a	36. a	46. c	56. d
7. a	17. c	27. b	37. a	47. d	57. b
8. d	18. a	28. d	38. d	48. d	
9. d	19. b	29. c	39. a	49. a	
10. b	20. d	30. c	40. b	50. c	

Poisons

Chemical Toxicology

KEY TERMS

Ames test

anticarcinogens

carcinogens

corrosive wastes

flammable wastes

hazardous waste

reactive wastes

teratogens

toxicology

toxic wastes

CHAPTER SUMMARY

20.1. All Things Are Poisons
   A. Toxicity depends on the chemical nature of the substance.
      1. A little salt is beneficial. It takes a lot of salt to kill an adult human.
      2. A few micrograms of a nerve poison can kill.
   B. Toxicity is different for different people.
      1. Sugar can kill a person with diabetes, but it is quite safe for others.
      2. Salt is dangerous for a person with hypertension, but quite safe for others.
   C. Toxicity depends on the route of administration.
      1. Nicotine is 50 times as toxic when injected as when taken orally.
      2. Water is fine in the stomach but deadly in the lungs (50 mL or so can cause drowning).
   D. Toxicity can vary greatly from one species of animal to another, making animal testing for toxicity suspect in some cases.
   E. Poisons Around the House
      1. Common household products are poisonous.
         a. Drain cleaners, oven cleaners, toilet-bowl cleaners
   F. Poisons in the Garden
      1. Herbicides and insecticides are poisonous
      2. Some plants are poisonous.
         a. Iris, azaleas, hydrangeas
         b. Holly berries, wisteria seeds, and privet hedge berries and leaves

20.2   Corrosive Poisons: A Closer Look
A. Acids and bases.
    1. Even in dilute solutions, acids and bases can hydrolyze proteins, destroying their function.
    2. Acids break down lung tissue.
B. Oxidizing Agents
    1. Ozone and other oxidizing agents deactivate enzymes by oxidizing sulfur-containing groups.
    2. The amino acid tryptophan undergoes a ring-opening oxidation with ozone.

20.3   Poisons Affecting Oxygen Transport and Oxidative Processes
A. Carbon monoxide bonds tightly to the iron atom in hemoglobin and hinders the transport of oxygen.
B. Nitrates are reduced to nitrites by bacteria in the digestive tract.
    1. Nitrites oxidize the iron in hemoglobin from Fe^{2+} to Fe^{3+}. The resulting methemoglobin cannot transport oxygen.
        a. In infants this condition is called "blue baby syndrome."
C. Cyanide: Agent of Death
    1. Cyanide, as HCN or its salts, is lethal in amounts of 50-60 mg.
    2. Cyanide binds cytochrome oxidases, tying up electrons and preventing their use for reduction reactions in the cell.
        a. Respiration ceases, and the cell dies quickly.
        b. If treated quickly, cyanide poisoning can be halted by administration of thiosulfate solutions, which convert cyanide ions to thiocyanate ions.
    3. Some people speculate that life on Earth arose from molecules formed by the polymerization of HCN.

20.4   Make Your Own Poison: Fluoroacetic Acid
A. When ingested, fluoroacetic acid is incorporated into fluorocitric acid.
    1. Fluorocitric acid blocks the energy-producing citric acid cycle by tying up the enzyme that acts on citric acid, energy production ceases, and the cell dies.
B. Sodium fluoroacetate (Compound 1080) was used both as a rat poison and to poison predators such as coyotes.
    1. Its use is banned on federal land due to the devastating effect it had on the eagle population.
    2. Sodium fluoroacetate is found in the South African gifblaar plant and is used by the natives to poison arrow tips.

20.5   Heavy Metal Poisons
A. Iron (as Fe^{2+} ions) is an essential nutrient.
    1. Too little leads to deficiency (anemia).
    2. Too much can be fatal, especially to a small child.
B. Heavy metals—those near the bottom of the periodic table—inactivate enzymes by reacting with sulfhydryl (SH) groups on enzymes.
    1. Arsenic compounds act in a similar manner (even though arsenic is not a metal).
C. Quicksilver →Slow Death
    1. Mercury metal is a liquid at room temperature; its vapor is especially hazardous.
    2. Mercury is a cumulative poison with a half-life in the body of 70 days.
    3. British Anti-Lewisite (BAL) is an antidote for mercury poisoning.
        a. It acts by chelating (tying up) mercury ions thus preventing them from attacking enzymes.
        b. BAL antidote is only effective when used right away.
D. Lead in the Environment
    1. Lead, a soft, dense metal that is corrosion resistant, has many uses.
    2. We get lead in foods, drinking water, and the air.
    3. Our bodies can excrete about 2 mg of lead per day.

4. Children with Pica Syndrome eat many things including chips of lead-based paint. They also get lead poisoning from playing in or near streets contaminated with lead from automobile exhausts.
   a. Lead poisoning in children causes mental retardation and neurological damage.
5. Lead poisoning is treated with BAL and ethylenediaminetetraacetic acid (EDTA).
   a. The calcium salt of EDTA replaces the lead ions with calcium ions and the lead-EDTA complex is then excreted.
   b. Damage to the nervous system is largely irreversible.
E. Cadmium: The "Ouch-Ouch" Disease
   1. Cadmium is used in alloys, electronics, and rechargeable batteries.
   2. In cadmiun poisoning cadmium ions substitute for calcium ions; this leads to a loss of calcium from bones, leaving them soft and easily broken.
   3. Cadmium poisoning from drinking water contamination in Japan led to a strange, painful malady known as itai-itai, the "ouch-ouch" disease.

20.6 More Chemistry of the Nervous System
A. Acetylcholine is a neurotransmitter; it carries messages across the synapse between cells.
B. After carrying its message, acetylcholine is broken down by the enzyme cholinesterase to acetic acid and choline.
   1. Acetylase then converts acetic acid and choline back to acetylcholine, completing a cycle.
C. Nerve Poisons: Stopping the Acetycholine Cycle
   1. Various chemical substances affect the acetylcholine cycle at different points.
      a. Botulin blocks the synthesis of acetylcholine.
         i. No messenger is formed, no message sent.
      b. Curare, atropine, and some local anesthetics block the receptor sites.
         i. The message is sent, but not received.
      c. Anticholinesterase poisons block the action of cholinesterase. This blocks the breakdown of acetylcholine; the nerves fire wildly and repeatedly, causing convulsions and death.
   2. Organic phosphates: insecticides and weapons of war.
      a. Organic phosphate nerve gases are very toxic and kill by inhalation or absorption through the skin.
      b. Chemical warfare agents are Tabun (Agent GA), Sarin (Agent GB), and Soman (Agent GD).
         i. The antidote for nerve gas is an atrophine injection.
      c. The insecticides malathion and parathion are closely related to the warfare agents.
   3. Nerve poisons have helped us to gain a better understanding of the nervous system.

20.7 The Lethal Dose
A. LD_{50}: A Measure of Toxicity
   1. Dose that kills 50% of a population of test animals.
   2. The larger the LD_{50} value, the less toxic the substance.

20.8 Your Liver: A Detox Tank
A. The liver is able to detoxify some poisons by
   1. Oxidation.
      a. Ethanol is oxidized to acetaldehyde, then to acetic acid, and finally to carbon dioxide and water.
      b. Nicotine is oxidized to less toxic cotinine.
   2. P-450 enzyme system can oxidize fat-soluble substances into water soluble ones that can be excreted.
      a. They can also conjugate compounds with amino acids.
         i. Toluene is oxidized to benzoic acid, which is conjugated with glycine to form hippuric acid, which is excreted.

B. Oxidation, reduction, and conjugation don't always detoxify.
1. Methanol is oxidized to more toxic formaldehyde.
2. Liver enzymes, built up through years of heavy ethanol use, deactivate the male hormone testosterone, leading to alcoholic impotence.
3. Benzene is oxidized to an epoxide that attacks key proteins.
4. Carbon tetrachloride is converted to trichloromethyl free radical that can trigger cancer.

20.9 Chemical Carcinogens: The Slow Poisons
A. Carcinogens cause the growth of tumors (abnormal growth of new tissue).
1. Benign tumors are characterized by slow growth; they do not invade new tissue.
2. Malignant tumors (cancers) generally grow irreversibly and invade and destroy other tissues.
B. What Causes Cancer?
1. Perhaps 80-90% are caused by environmental factors..
a. Of this, about 40% are caused by cigarette smoking.
b. Dietary factors (high-fat, low-fiber foods, pickled foods, charcoal-broiled foods) cause about 25-30%.
c. About 10% are caused by occupational exposure (to asbestos, benzene, vinyl chloride, coal tars, etc.).
d. Perhaps 10-15% are caused by environmental pollutants.
2. Perhaps 10-20% are caused by genetic factors and viruses.
C. There are many natural carcinogens (aflatoxins, safrole).
D. How Do Cancers Develop?
1. Some carcinogens modify the DNA, scrambling the code for replication and protein synthesis.
2. Carcinogens, radiation, and some viruses activate oncogenes, which regulate cell growth and division. In addition, suppressor genes must be inactivated before a cancer can develop.
E. Chemical carcinogens
1. Polycyclic aromatic hydrocarbons, such as 3,4-benzpyrene.
a. Found in grilled meat, coffee, and cigarette smoke.
2. Aromatic amines (ß-naphthylamine, benzidine).
a. Used in the dye industry.
3. Aminoazo dyes (4-dimethylaminoazobenzene or "butter yellow").
a. Previously used to color margarine.
4. Aliphatic compounds (dimethylnitrosamines, vinyl chloride.
F. Anticarcinogens are found in food.
1. They include fiber, antioxidant vitamins (Vitamins C, E and beta-carotene—precursor to Vitamin A), and cruciferous vegetables.

20.10 Testing for Carcinogens: Three Ways
A. The Ames test is used to test chemicals to see if they cause mutations in bacteria. Mutagens are often carcinogens.
B. Animal testing
1. Animal tests employ large doses of the suspected substance on small numbers of laboratory animals such as rats.
a. Such tests have limited relevance; humans are exposed to lower doses for longer periods, and human metabolism can be different from that of the experimental animals.
b. However, good correlation exists between known human carcinogens and cancer in laboratory animals.
C. Epidemiological studies correlate the incidence of various cancers in human populations with various work, dietary practices, and with exposure to chemicals.

20.11 Birth Defects: Teratogens
A. Teratogens are chemical substances that cause birth defects.

1. Thalidomide was a popular tranquilizer and morning sickness medicine for pregnant women in Germany and Great Britain during the 1950s and 1960s.
2. Thalidomide was responsible for numerous birth defects and is now banned.
3. Thalidomide was never approved for use in the United States.

20.12 Hazardous Wastes
    A. Hazardous wastes are those that can cause or contribute to death or illness or that threaten human health or the environment when improperly managed.
    B. There are five classes of hazardous wastes (products that cause or contribute to death or human illness).
        1. <u>Reactive</u> <u>wastes</u> react spontaneously.
            a. They react violently with air or water.
            b. They explode when exposed to shock or heat.
                i. Examples of reactive wastes are sodium metal, TNT, and nitroglycerin.
        2. <u>Flammable</u> <u>wastes</u> are those that burn readily upon ignition.
            a. Hexane is an example of a flammable waste.
        3. <u>Toxic</u> <u>wastes</u> contain or release toxic substances in quantities that pose a hazard to human health or the environment.
            a. Examples of toxic wastes include polychlorinated biphenyls (PCBs).
        4. <u>Corrosive</u> <u>wastes</u> corrode their containers.
            a. Examples of corrosive wastes include strong acids and bases.
        5. <u>Organic</u> <u>hazardous</u> <u>wastes</u> can be destroyed by incineration.

20.13 What Price Poisons?
    A. Many useful substances are toxic, especially when misused.
    B. We must balance the risks associated with a product against the benefits that we gain by using it.

CHAPTER OBJECTIVES

(You should...)

1. Know that strong acids, strong bases, and strong oxidizing agents are corrosive poisons.

2. Know that ozone can deactivate enzymes by oxidizing them, particularly those that include the sulfur-containing amino acids cysteine and methionine.

3. Know that nitrites oxidize the iron in hemoglobin from Fe^{2+} to Fe^{3+}. The resulting methemoglobin is incapable of transporting oxygen.

4. Know that cyanide blocks the oxidation of glucose in the cells by tying up iron-containing enzymes.

5. Know that thiosulfate is the antidote for cyanide poisoning. It acts by converting cyanide (CN^-) to relatively harmless thiocyanate (SCN^-).

6. Know that fluoroacetic acid is incorporated into flurocitric acid, which blocks the citric acid cycle shutting off energy production in the cell.

7. Know that iron, iodine, and other minerals, in moderate amounts, are necessary nutrients; in larger amounts, they are toxic.

8. Be able to recognize mercury, lead, and cadmium as heavy metal poisons.

9. Know that arsenic, a nonmetal, mimics the action of heavy metal poisons.

10. Know that arsenic and mercury poisoning are treated by intravenous injection of **BAL**.

11. Know that in cases of severe lead poisoning, the patient is treated with **BAL and EDTA** to remove lead from soft tissue.

12. Know that the body can rid itself of about half of its load of mercury in about 70 days.

13. Know that we can excrete about 2 mg of lead per day.

14. Know that cadmium poisoning leads to a loss of calcium from bones, leaving them soft and easily broken.

15. Know three ways the acetylcholine cycle can be interrupted by poisons. **Be able to list one poison of each type** and describe the symptoms of that type of poison.

16. Know that atropine is the antidote for organophosphorus poisoning.

17. Know that the liver oxidizes foreign substances. This often serves to detoxify compounds, but it may make some substances more toxic.

18. Know that the same liver enzymes that detoxify alcohol also deactivate testosterone, the male hormone. This is the mechanism for alcoholic impotence.

19. Know that there are three kinds of studies to indicate if a chemical compound is carcinogenic: (a) the Ames test, (b) animal studies, and (c) epidemiological studies.

20. Be able to recognize members of four major classes of carcinogens; aromatic amines, polycyclic aromatic hydrocarbons, aliphatic compounds, and aminoazo dyes.

21. Know that fiber, BHT, and ß-carotene are thought to have anticarcinogenic activity.

22. Know that diets high in cruciferous vegetables (cabbage, broccoli, brussels sprouts, cauliflower, and kale) have been shown to reduce the incidence of cancer in laboratory animals and in humans.

23. Know that teratogens are substances that cause birth defects.

24. Know that thalidomide, once used in Germany and Great Britain to treat morning sickness, is a potent teratogen.

25. Be able to define the term hazardous waste.

26. Know the five kinds of hazardous wastes.

27. Be able to give an example of each of the five kinds of hazardous wastes and a method of disposal for each.

DISCUSSION

"All things are poison." That isn't far from the truth, but some things are much more poisonous than others. Also, some present an immediate hazard (acute poisons), while others exhibit their harmful effect only over a much longer period of time (chronic poisons). The public often gets a voodoo cocktail of misinformation about toxic chemicals. A few chemical principles will help you to gain a better understanding of what toxic substances are, how they act, and what can be done about toxic wastes.

SELF-TEST

Multiple Choice

1. Sulfuric acid is a

    a. carcinogen
    c. corrosive poison

    b. nerve poison
    d. blister agent

2. Which of the following is a corrosive poison?

    a. carbon monoxide
    c. nicotine

    b. cyanide
    d. sodium hydroxide

3. The toxicity of a substance depends upon

    a. its chemical nature
    c. the person or animal subjected to it

    b. its route of administration
    d. all of these

4. Which of the following is a blood agent?

    a. carbon monoxide
    c. ozone

    b. nitric acid
    d. thiosulfate

5. Ozone can deactivate enzymes by

    a. chelating them
    b. oxidizing sulfur-containing amino acids
    c. reducing amino groups
    d. hydrolyzing peptide linkages

6. Nitrites convert hemoglobin to

    a. oxyhemoglobin
    c. nitrohemoglobin

    b. carboxyhemoglobin
    d. methemoglobin

7. Cyanide exerts its toxic effect by blocking

    a. oxidation of glucose in the cells
    c. oxygen transport

    b. the citric acid cycle
    d. the excretion of lead

8. Thiosulfate is the antidote for cyanide poisoning. It acts by converting cyanide to relatively harmless

       a.   carbon dioxide   b.   cyanate
       c.   nitrate   d.   thiocyanate

9. Fluoroacetic acid acts as a poison by blocking

       a.   the acetylcholine cycle   b.   the citric acid cycle
       c.   oxygen transport   d.   tooth decay

10. Iron (as Fe^{2+}) is
       a.   toxic in any amount
       b.   safe in any amount
       c.   toxic in large amounts but necessary in smaller amounts
       d.   toxic in large amounts but unnecessary in human nutrition

11. Which of the following is a heavy metal poison?

       a.   iodine   b.   lead
       c.   lithium   d.   sodium

12. Which nonmetal mimics the action of a heavy metal poison?

       a.   arsenic   b   chlorine
       c.   fluorine   d.   sulfur

13. Little Lavinia visits the lab and swallows some mercury. A test shows that her blood has 10 ppm mercury. Her load will be down to 5 ppm in about

       a.   48 hours   b.   7 days
       c.   70 days   d.   7 years

14. Which kind of poisoning results in brittle, easily broken bones?

       a.   cadmium   b.   cyanide
       c.   lead   d.   ozone

15. Arsenic poisoning is treated by intravenous injection of

       a.   cyanide   b.   atropine
       c.   BAL   d.   thiosulfate

16. Which substance is a neurotransmitter?

       a.   acetylcholine   b.   atropine
       c.   ß-naphthylamine   d.   parathion

17. Little Lavinia eats some mushroom soup contaminated with botulin. She is paralyzed because nerve messages

    a. are never sent   b. are sent but not received
    c. are received continuously   d. are garbled

18. Funky drinks an ounce of organophosphorus pesticide. To save his life, Lavinia should inject

    a. EDTA   b. BAL
    c. atropine   d. thiosulfate

19. Atropine, the antidote for nerve gas poisoning, acts by blocking which part of the acetylcholine cycle?

    a. breakdown   b. formation
    c. release   d. receptors

20. Little Lavinia lives it up too many times at the Tipplers Tavern. The alcohol she drinks is oxidized in her oversized

    a. tongue   b. stomach
    c. liver   d. kidneys

21. Liver enzymes, built up to deactivate alcohol over years of heavy drinking, render the alcoholic impotent by deactivating

    a. testosterone   b. serotonin
    c. danazol   d. aphrodisiacs

22. Which substance is not detoxified by the liver?

    a. ethanol   b. nicotine
    c. parathion   d. toluene

23. Butter yellow is a member of which class of carcinogens?

    a. aromatic amines   b. aminoazo dyes
    c. polycyclic aromatic hydrocarbons   d. aliphatic compounds

24. Benzpyrene is a member of which class of carcinogens?

    a. aromatic amines   b. aminoazo dyes
    c. polycyclic aromatic hydrocarbons   d. aliphatic compounds

25. β-Naphthylamine is a member of which class of carcinogens?

    a. aromatic amines   b. aminoazo dyes
    c. polycyclic aromatic hydrocarbons   d. aliphatic compounds

26. Which of the following is known to be a carcinogen?

    a. aflatoxin   b. ß-carotene
    c. ozone   d. penicillamine

27. The principal cause of cancer is

    a. cigarette smoking   b. environmental pollution
    c. food additives   d. occupational exposure to carcinogens

28. Which substance is thought to have anticarcinogenic activity?

    a. BHT   b. hexane
    c. penicillamine   d. thalidomide

29. We can reduce our chances of getting cancer by eating a lot of

    a. cabbage   b. eggs
    c. peas   d. potato chips

30. Thalidomide is

    a. an acute poison   b. a carcinogen
    c. a mutagen   d. a teratogen

31. An example of a reactive hazardous waste is

    a. benzene   b. dichloromethane
    c. mercury   d. sodium

32. An example of a flammable hazardous waste is

    a. acetone   b. arsenic
    c. dichloromethane   d. lead

33. An example of a toxic waste is

    a. ethanol   b. hexane
    c. potassium   d. sodium cyanide

34. Which class of wastes cannot be destroyed by incineration?

    a. chlorinated hydrocarbons   b. heavy metals
    c. hydrocarbons   d. sewage solids

35. Poisons are found

    a. only in chemistry labs   b. only in hardware stores
    c. only in the home   d. everywhere

36. Corrosive poisons act by

    a.   breaking down fats   b.   breaking down water
    c.   breaking down proteins   d.   breaking down electrolytes

37. Oxidizing poisons

    a.   cause fires   b.   put out fires
    c.   break chemical bonds   d.   are specific in their activity

38. Carbon monoxide

    a.   blocks the transport of oxygen   b.   oxidizes to carbon dioxide
    c.   reduces to carbon   d.   all of the above

39. Heavy metals are dangerous

    a.   in all amounts   b.   only above certain levels
    c.   only below certain levels   d.   in amounts specific to each metal

40. Who is most susceptible to poisons?

    a.   Little Lavinia   b.   Medium Lavinia
    c.   Big Lavinia   d.   all are equally at risk

41. Most nerve poisons act by

    a.   blocking receptor sites   b.   opening receptor sites
    c.   destroying nerve cells   d.   all of the above

42. Which organ naturally "cleans" the body of toxins?

    a.   heart   b.   lungs
    c.   liver   d.   kidney

43. An LD_{50} means

    a.   50% die   b.   50 g of poison is toxic
    c.   those over 50 years of age die   d.   all of the above

44. We have identified _____ human carcinogens

    a.   3
    b.   30
    c.   3,000
    d.   3,000,000

45. The Ames test is based upon

    a.   epidemiology study of Ames, Iowa
    b.   that carcinogens are usually acid
    c.   that carcinogens are usually mutagens
    d.   that carcinogens are usually chemicals

46. Teratogens cause

    a.   birth defects   b.   shortness of breath
    c.   cancer   d.   heart disease

ANSWERS

1. c	11. b	21. a	31. d	41. a
2. d	12. a	22. c	32. a	42. c
3. d	13. c	23. b	33. d	43. a
4. a	14. a	24. c	34. b	44. b
5. b	15. c	25. a	35. d	45. c
6. d	16. a	26. a	36. c	46. a
7. a	17. a	27. a	37. c	
8. d	18. c	28. a	38. a	
9. b	19. d	29. a	39. d	
10. c	20. c	30. d	40. a	